Meaningful Math in Preschool

Making Math Count Throughout the Day

Meaningful Math in Preschool

Making Math Count Throughout the Day

Polly Neill
with Suzanne Gainsley

Published by
HighScope® Press

A division of the
HighScope Educational Research Foundation
600 North River Street
Ypsilanti, Michigan 48198-2898
734.485.2000, FAX 734.485.0704

Orders: 800.40.PRESS; Fax: 800.442.4FAX; www.highscope.org
E-mail: press@highscope.org

Copyright © 2014 by HighScope Educational Research Foundation. All rights reserved. Except as permitted under the Copyright Act of 1976, no part of this book may be reproduced or distributed in any form or by any means, electronic or mechanical, including photocopy, recording, or any information storage-and-retrieval system, without either the prior written permission from the publisher, or authorization through payment of the appropriate per-copy fee to the Copyright Clearance Center, Inc., 222 Rosewood Drive, Danvers, MA 01923, 978.750.8400, fax 978.646.8600, or on the web at www.copyright.com. The name "HighScope" and its corporate logos are registered trademarks and service marks of the HighScope Foundation.

Editors: Joanne Tangorra, Marcella Fecteau Weiner
Cover design: Phire Advertising and Design LLC
Text design and production: Judy Seling, Seling Design LLC
Photography: All photos by Bob Foran, Ann Arbor, MI, with the following exceptions:
Gregory Fox — 7, 8, 12, 19 (top, bottom), 21, 23, 44, 50, 63, 67, 69, 72, 74, 80, 85, 95, 98, 111, 117, 121, 130, 132, 133, 135, 137, 139, 147
HighScope staff — Front cover, 19 (middle), 34, 36, 39, 47, 75, 90, 99, 120, 142
Pat Thompson — 26, 55, 123

Library of Congress Cataloging-in-Publication Data
Neill, Polly, author.
  Meaningful math in preschool : making math count throughout the day / Polly Neill with Suzanne Gainsley.
       pages cm
  Includes bibliographical references.
  ISBN 978-1-57379-678-1 (soft cover : alk. paper) 1. Mathematics--Study and teaching (Early childhood)--Activity programs. I. Gainsley, Suzanne, 1964- author. II. Title.
  QA135.6.N45 2014
  372.7'049--dc23
  2014017154

Printed in the United States of America
10 9 8 7 6 5 4

Dedication

To my big sister, Allen Neill Schauffler, my early childhood hero, who introduced me to this place called HighScope and whose only flaw is the "thinking step."

Contents

Acknowledgments xi

Chapter 1. The Importance of Early Childhood Mathematics 1
    The Mathematics Gap 2
    Math Doesn't Just "Happen" 4
    Teaching Early Mathematics With Confidence 6
    Rediscovering the Fun in Math 8
    How to Use This Book 11

Chapter 2. The HighScope Approach to Math 13
    Key Developmental Indicators (KDIs) 13
    Math Throughout the Daily Routine 14

Chapter 3. Creating an Environment That Invites Math Learning 25
    Designing a Math-Friendly Environment 25
    Block Area 26
    Block Area Materials 30
    House Area 34
    House Area Materials 35
    Toy Area 38
    Toy Area Materials 38
    Art Area 41
    Art Area Materials 42
    Sand and Water Area 43
    Sand and Water Area Materials 43
    Outside Area 45
    Outside Area Materials 45

Chapter 4. Interacting With Children to Support Math Learning 49
    Child Language and Thought 49
    Adult-Child Interactions 51
    Strategies to Support Active Learning 52
    Adult Scaffolding 59

Chapter 5. Greeting Time 61
    What Children and Adults Do During Greeting Time 61
    Over There! Catch the Math During Greeting Time 62
    Math in Action: Waiting for Lindsay 65

Chapter 6. Planning and Recall Times 71
    What Children and Adults Do During Planning Time 71
    Over There! Catch the Math During Planning Time 72
    Math in Action: Planning With Sammy and Claire 74

What Children and Adults Do During Recall Time 76
Over There! Catch the Math During Recall Time 76
Math in Action: Using a Recall Chart 79

Chapter 7. Work Time 81
What Children and Adults Do During Work Time 81
Over There! Catch the Math in the Art Area 82
Math in Action: Brianna's Collage 84
Over There! Catch the Math in the Block Area 86
Math in Action: Razi and Henry Build a Road 88
Over There! Catch the Math in the House Area 89
Math in Action: Jamee Has a Checkup 91
Over There! Catch the Math in the Sand and Water Area 91
Math in Action: Sand and Water Area 93
Over There! Catch the Math in the Toy Area 97
Math in Action: Amare Plays With Inch Blocks and Small Reptiles 98
Math in Action: Kyra, Magna-Tiles, and Peach Pie 99
Math in Action: Diego and Avery Weigh Shells and Stones 100
Over There! Catch the Math in the Book Area 101
Math in Action: Randy Reads *The Very Hungry Caterpillar* 102
Math in Action: Across the Days and Throughout the Classroom 102

Chapter 8. Small-Group Time 105
What Children and Adults Do During Small-Group Time 105
Over There! Catch the Math During Small-Group Time 106
Math in Action: Golf Tees and Styrofoam 109
Math in Action: Shape Blocks 110

Chapter 9. Large-Group Time 115
What Children and Adults Do During Large-Group Time 115
Over There! Catch the Math During Large-Group Time 116
Math in Action: *The Three Billy Goats Gruff* 118
Math in Action: Let's Have a Parade! 120
Math in Action: Five in the Bed 122

Chapter 10. Cleanup Time, Mealtimes, and Transitions 125
Cleanup Time 125
What Children and Adults Do During Cleanup Time 125
Over There! Catch the Math During Cleanup Time 127
Math in Action: Cleanup Tickets 128
Mealtimes 130
What Children and Adults Do During Mealtimes 131
Over There! Catch the Math During Mealtimes 131
Math in Action: Cheese and Crackers for Snack 133

Transitions 134
What Children and Adults Do During Transitions 134
Over There! Catch the Math During Transitions 136
Math in Action: Going to the Lunchroom 137

Chapter 11. Outside Time 139
What Children and Adults Do During Outside Time 139
Over There! Catch the Math During Outside Time 140
Math in Action: Kicking Balls With Evan and Sandy 143
Math in Action: Brynna and Jordyn Get Speeding Tickets 144
Math in Action: Turns on the Swing 145

Appendix A: HighScope Key Developmental Indicators 150

Appendix B: HighScope Daily Routine 152

Appendix C: HighScope Preschool Wheel of Learning 153

References 155

Acknowledgments

The important thing to remember about mathematics is not to be frightened.
— Dawkins (1996, p. 67)

I am profoundly grateful to several people who could continually be called on for clarification, modifications, and support. Ann S. Epstein, Beth Marshall, and Sue Gainsley, the team that refined and revised HighScope's math curriculum, were invaluable, incredibly patient, and generous with their time and encouragement. My colleagues in the Early Childhood Department, Shannon Lockhart and Kay Rush, provided their rich knowledge and experience with children plus unfailing support. Other colleagues who also are (or were) the talented and remarkable teachers in the HighScope Demonstration Preschool — Christine Snyder, Becky James, Molly Jourden, and Emily Thompson — offered rich examples that I could draw from. Linda Horne, our department's administrative assistant, ensured that everything proceeded with efficiency, cooperation, and humor.

That this book is well organized, consistent, and clear is all due to Joanne Tangorra, talented writer and expert editor, who kept asking me the tough questions. Her colleague, editor Marcella Fecteau Weiner, came in as the "ace closer" and added final polish through the book's final design and printing stage.

This book could not have been completed without the contributions of Sue Gainsley. Sue's years of classroom experience and her intuitive sense for what is right for children resulted in realistic scenes that teachers will recognize and identify with. The longer scenarios and support strategies in each chapter are the result of Sue's talented writing as well as her insightful classroom observations and provide valuable learning experiences.

Finally, Mr. Goudey was one of *those* teachers; he taught my high school Algebra 2 class with skill, humor, and incredible patience. It is because of these qualities and his willingness to spend extra hours explaining some of the finer points of algebra that I have been able to apply mathematics throughout my university and professional life.

The Importance of Early Childhood Mathematics

One and one and one is three.
— The Beatles (1969/2009, track 1)

At greeting time, Claire writes her name and then shows Shannon (an adult). Claire says, "I wrote my name backwards."

▲

When Delilah (a teacher) suggests reversing the numbers on the message board, Desmond says to her, "If you do that (reverse the 1–2–3 order of the messages), the 3 would be on top, the 2 in the middle, and the 1 on the bottom."

▲

At snacktime, Evan looks around the table, points to each empty chair, and says, "One missing, one missing, one missing." Rhonda (an adult) points to each chair and says, "One, two…" and then Evan adds, "Three!"

▲

After she listens to a story about a child that sleeps in a crib, Faith says, "I sleep in a grown-up bed." Amani (an adult) asks her what she means by "grown-up." Faith says, "That means you have a big bed that is bigger than a medium bed."

Being a witness to a developing child's mind is a privilege and an adventure. As the previous anecdotes suggest, children are natural mathematicians — curious, open, and adventurous. They "have unique ways of describing phenomena and using their imagination to explore their own ideas. All this experience adds up to a personal sense of mathematics" (Gross, 2006, p. 47). Similarly, Eugene Geist (2009), author of *Children Are Born Mathematicians*, suggests that we "begin to think of children as competent mathematicians who, while working on age-appropriate problems, are using the same thought processes as advanced mathematicians" (p. 12).

While educators and researchers have long agreed that children's literacy learning begins the day they are born, they have only recently come to realize that mathematics learning starts just as early in a child's development. The early childhood professional community now recognizes that from birth, children begin to construct a "foundation for future mathematical concepts" (Geist, 2009, p. 5) as they learn about their world — through their physical, social, and mental interactions with their environment and the people in it.

In fact, the findings of researchers in fields from neuropsychology to human ecology have completely revised the traditional view that young children are capable of little more than simple rote counting, perhaps to 10. Researchers' thinking has evolved from a belief that young children did not (and could not) understand mathematics to an understanding that even the youngest children have innate mathematical abilities (Clements, 2001). An article in *The New Yorker*, for example, describes the pioneering research of neuroscientist Stanislas Dehaene, whose work suggests that "we are all born with an evolutionarily ancient mathematical instinct" (Holt, 2008, p. 2). Simply put: Infants' brains are wired with the ability to sense and represent number.

According to a 2010 joint position statement of the National Association for the Education of Young Children (NAEYC) and the National Council of Teachers of Mathematics (NCTM), research indicates that children's early math experiences have "long-lasting outcomes" (p. 2). For example, an analysis of six longitudinal data sets relating school-entry skills to later teacher ratings and test scores of reading and math achievement (Duncan et al., 2007) offered some surprising results:

- "Early math concepts, [such] as knowledge of numbers and ordinality (e.g., first, second, third), were the most powerful predictors of later learning.
- Early math is a more powerful predictor of later reading achievement than early reading is of later math achievement" (p. 1443).

In its report *Mathematics Learning in Early Childhood: Paths Toward Excellence and Equity,* the National Research Council (NRC; 2009) also stresses the importance of mathematics in other skills and areas of learning: "Academic activities such as mathematics can be a context in which social-emotional development and the foundations of language and literacy flourish" (p. 2).

The Mathematics Gap

While educators of young children have increasingly come to recognize the importance of mathematics learning to children's success in school and society, assessments of math proficiency among students indicate that "US children's mathematical proficiency is far below that of many other countries, and the mathematics gap is wider for children living in poverty and those who are members of ethnic, cultural, and linguistic minority groups" (NAEYC & NCTM, 2009, p. 1). The NRC (2009) also expresses "particular concern about the chronically low mathematics… performance of economically disadvantaged students and the lack of diversity in the science and technical workforce" (p. 1) and cites research findings that "reveal enormous discrepancies in young children's levels of mathematics competence" (p. 95).

That there is a critical need for high-quality mathematics for all children, especially those living in poverty, has been widely documented. A study on pre–K mathematics intervention noted the learning environments of children from middle-income families are more rich mathematically than the learning environments of children from low-income families (Klein, Starkey, Molfese, Brown, & Molfese, 2008). Children from low-income families "begin school with much less mathematical knowledge than their wealthier peers," a disparity related to the fact that low-income parents engage in fewer and

Recent research shows that early math is a better predictor of later reading skills than early reading is of later math skills.

less frequent mathematical activities with their children at home, compared to more affluent parents (Siegler, 2009, p. 118). As noted earlier, infants everywhere are born with a preverbal number sense. However, subsequent number sense development, such as understanding number words and symbols, is strongly affected by intentional experiences and activities. The difference in mathematical knowledge between children from low-income families and children from higher-income families is evident in their basic mathematic skills, including counting, recognizing numbers, adding and subtracting, and measuring (Siegler, 2009).

With clear evidence that a math gap exists, the early childhood field is considering the long-term implications for those children at risk for being left behind. How do we take the necessary steps to close that gap to ensure academic success for all children? As Siegler (2009) points out, "children's mathematical knowledge in kindergarten predicts their math achievement test scores in elementary school, middle school, and even high school" (p. 118). Moreover, Siegler reports the connection that exists between early and later mathematical knowledge is approximately twice as strong as the one between early and later reading achievement.

The place to start addressing the mathematics achievement gap is early childhood, where every educational setting should provide *all* children with *"high-quality, challenging,* and *accessible* mathematics experiences" (NAEYC & NCTM, 2009, p. 1). Following are several of the recommendations made by the NRC in its 2009 report that are particularly pertinent to the issues addressed in this book. At the top of the NRC's list of recommendations is one that addresses early childhood mathematics at the national level: "Recommendation 1. A coordinated national early childhood mathematics initiative should be put in place to improve mathematics teaching and learning for all children ages 3 to 6" (p. 345).

Another NRC recommendation specifies the most important math concepts to focus on: "Recommendation 2. Mathematics experiences in early childhood settings should concentrate on (1) number (which includes whole number, operations, and relations) and (2) geometry, spatial relations, and measurement, with more mathematics learning time devoted to number than to other topics" (p. 345). Additional NRC recommendations address the need for professional development for early childhood teachers that will help them understand the essential areas of mathematics for early learners, the way certain areas of mathematics develop, and the principles of intentional teaching.

Adults in the preschool classroom can encourage and support young children's mathematics learning by supporting and extending their existing mathematics knowledge in everyday situations. For example, a child mathematizes when she sorts her snack mix into its different ingredients and then says, "I have five more goldfish than pretzels"; or sets the table for a birthday party for six children and then, realizing she also invited the teachers, says, "I need two more plates and forks for Mariah and Seth"; or puts her name on the computer sign-up sheet and says, "I'm third, after Joey; he's second." Mathematizing is "a means for children to deepen, extend, elaborate and refine their thinking as they explore ideas and lines of reasoning" (Fuson, Clements, & Beckmann, 2010, p. 2). As they *mathematize,* children "solve problems;…reason and communicate their reasoning;…represent ideas using objects, drawings, written symbols, or internal visualization; and…connect ideas" (p. 3).

Math Doesn't Just "Happen"

As the field of early childhood has gradually come to understand that young children have "the capacity and interest to learn meaningful mathematics" (NRC, 2009, p. 333) and that they do so through everyday experiences as early as their first year of life, we are also coming to understand how we can best build on and extend their early math learning. While it is certainly true that young children use math every day without realizing it — in routine play and activities from figuring out who jumped higher to identifying shapes on street signs — this does not mean that math learning "can be haphazard or left to chance" (Epstein, 2012, p. 6). As the NRC (2009) states, "emerging research indicates… that learning experiences in which mathematics is a supplementary activity rather than the primary focus are less effective in promoting children's learning than experiences in which mathematics is the primary goal" (p. 2). In other words, adults in early childhood programs must intentionally encourage and extend children's mathematical understanding "by providing appropriate materials and encouraging reflective thinking to help preschoolers construct mathematical knowledge from their experiences" (Epstein, 2012, p. 14).

Eugene Geist (2009) puts forth a set of three principles for using best practices to teach mathematics:

"Thinking about the problem, not the answer, is what is most important" (pp. 2–3). In several of the countries whose students surpass those in the United States in mathematics scores, children are taught to use math tools, that is, the tools necessary to help them figure out a problem, not necessarily find the right answer. By contrast, the tradition in the United States has been an emphasis on memorization of formulas and practice, with the sole focus on getting the right answer. According to Geist, children whose math scores surpass US children have to come to understand that the "correct answers do not come from a teacher" (p. 3) but from learning to think for themselves.

"Process is more important than product" (pp. 3–4). This expression is nearly a cliché among early childhood professionals.

To help children develop math skills, adults must provide them with appropriate materials and support them while they construct their own mathematical knowledge.

However, in mathematics, it means quite literally that math is much more than just the facts (the answer/product). What is of greater importance are the processes (the tools) that serve us as we figure out problems. An unsuccessful result (i.e., a wrong answer) is just another step on the way to solving the problem. As Geist explains, "if children are directly taught addition 'facts' before they personally experience addition through play and manipulation, they will learn a set of unrelated facts such as '5 + 4 = 9' with no conceptual understanding of number to link them to" (p. 3). Remember, mathematicians don't compete for speed; they take their time on problems.

"Answers come from a logical certainty, not from an authority figure" (p. 4). When people write about the development of infants, toddlers, and preschoolers, they often begin by making a statement such as "children are born explorers (or scientists, problem solvers, adventurers, and so on)." This is true. Infants and toddlers, curious from the start, investigate their environment with passion and persistence. As preschoolers refine the ability to represent objects, persons, and events that are not present (otherwise known as symbolic thought), they become more adept at solving problems. With its increasing use, "symbolic thought introduces the ability to use numbers to represent quantities of objects" (p. 4). For example, when Jacob is asked how old he is, he holds up three fingers and says, "I'm three."

Teaching Early Mathematics With Confidence

As teachers of young children, it is exciting to know that preschoolers have a natural interest and inclination to learn mathematics. In fact, when it comes to math, children "are often more confident and adventurous than many adults for whom 'math anxiety' was an unfortunate part of growing up" (Epstein, 2012, p. 2).

Somewhere in our mathematics education many of us lost the sense of curiosity and delight in mathematical concepts that we had as young children. Remember eating your M&Ms in a particular order; measuring out the flour for cookies with your big sister; or building elaborate constructions in the basement incorporating blocks, cars, pillows, blankets, orange crates, and even a couple of old two-by-fours filched from Dad's shop? For many of us, something changed when we passed through the elementary school doors. We were led to believe that we couldn't learn math by grouping candy or cereal, making cookies, or building forts. The only way to learn math was to do so in isolation from real materials, accompanied by never-ending memorization, work sheets, flashcards, and recitation, rather than by thinking, puzzling, speculating, experimenting, collaborating, and repeated trial and error with real materials.

What were your math learning experiences like? Remember those word problems that began like this: One train leaves the station at 1 p.m. travelling at 40 mph.... Or questions like this: If $x + 2x$ is 5 more than $y + 2y$, then $x - y =$ ____? How would you feel about solving one of those problems today? Confident? A bit rusty? Wishing you had paid attention? Or, perhaps like me, just the phrase, "One train leaves the station..." still produces vivid feelings of fear and self-doubt.

In "A Study of Mathematics Anxiety in Pre-Service Teachers," Gina Gresham (2007) defines mathematics anxiety "as a feeling of helplessness, tension, or panic when asked to perform mathematics or problems" (p. 182). Unfortunately, it not only affects the people who suffer from it, but also may result in poor academic performance of their future students. Gresham cites research that shows a "disproportionately large percentage of pre-service teachers experience significantly high levels of mathematics anxiety" (p. 183). This, in turn, undermines their confidence in their ability to teach children mathematics. Furthering the cycle, the teachers' dislike of, and discomfort with, mathematics may, in turn, be transmitted to their students. Researchers have found that teachers often repeat how they were taught math — including rote memorization tasks in which errors are regarded as mistakes — rather than teaching math as an opportunity to experiment to see what works or doesn't work (Gresham, 2007; Hynes-Berry & Itzkowich, 2009). With such strong feelings remaining so many years after we may have struggled with math, it is not surprising that so many of us ask ourselves, "How can I possibly teach these children math with confidence, let alone relay a sense of excitement and joy about the process?" Remember, adults like children, use mathematics every day, often without realizing it. The following scenario may provide some reassurance that you, too, can overcome your math anxiety and teach children with confidence:

> You're at the grocery store, cart loaded, and you're ready to check out. What next? You glance at each of the checkout lines, noting which is the longest. You also look to see how full their carts are and which checkout stands have baggers and which don't. You may even notice the speed of the cashiers. In less than a minute, you have taken in a considerable amount of (mathematical) information, estimate in your head the effects of all those variables, and choose the line you think will move fastest. This is not just simple mathematics; this is sophisticated and complex

What Ever Happened to Arithmetic?

Many of us did not learn "math" until we reached junior high; up to that point, we were taught "arithmetic." Today children learn math right from the start. Has mathematics replaced arithmetic in education? And, if so, what is the difference between them? To answer these questions, let us be certain we are clear as to what we are talking about. The Merriam-Webster Dictionary simply defines *arithmetic* as a "branch of mathematics that deals with numbers and their addition, subtraction, multiplication, and division" ("Arithmetic," 2013).

So what happened to arithmetic in the classroom? Nothing really "happened" to arithmetic. It is part of mathematics and is still a subject children study. However, the word *mathematics* is used in education today to emphasize that there is much more to this broad topic than basic arithmetic operations. Arithmetic is just one part of mathematics — and an important part — but still just a part of a broader subject that also encompasses the science of number, quantity, and space.

A child may not be successful in solving a problem the first, second, or even ninth time, but these should not be considered mistakes. Rather, they should be viewed as attempts to solve a problem in a way that makes sense to that child.

mathematics that includes solving systems of simultaneous equations and incorporating probability of error estimates. Yet you do it almost effortlessly. (Sorooshian, 2009, par. 3)

Rediscovering the Fun in Math

As we become adults, we develop filters that limit our ability to be "true learners," and we become too focused on the right answer (Hynes-Berry & Itzkowich, 2009). For us to think about mathematical concepts as children do (as true learners), we need to understand how children think and learn, which is different from how we think and learn.

Because children think about mathematics differently than adults, they often make what seem to be "mistakes" in logic. *Children, however, do not make mistakes;* they are simply making one more attempt at solving a problem (Hynes-Berry & Itzkowich, 2009). For example, preschoolers might say that a long row of five blocks has more than a shorter row of the same number; they are simply overgeneralizing an otherwise valid concept. If we take the time to understand young children's thinking, we can see how they solve problems in a way that is logical (i.e., makes sense) to them.

If adults are going to try being "new learners, they [have to] bypass their competence as adults and move closer to the consciousness of the young child who experiences learning in a more organic fashion through play" (Hynes-Berry & Itzkowich, 2009, p. 110). If we help teachers learn to interact with materials the way children do, then teachers are more likely to "recognize and promote how young children move through the landscape of learning, from confusion to misconception to understanding" (p. 109).

This is not to suggest that teachers act like children, but rather that they take some time (on their own) to rediscover and play with the materials in their classrooms in their own way. Teachers need to first use the materials themselves and fully experience their properties (e.g., How does clay feel as it dries over time? What are the challenges in making a car stay on the ramp?). Then they need to reflect on their child development knowledge and classroom observations (e.g., How do young children experience ordinality? What is a preschooler's understanding of spatial concepts such as *middle* and *edge*?). Only by combining hands-on experience with knowledge and reflection can teachers understand how children might interact with materials and draw conclusions in the context of their mathematical explorations.

Here are some examples, supported with reflective anecdotes, of how you, as a teacher, might become a "new learner":

Build in the block area with your coteacher. Set up some ramps and bridges, and then notice what happens when you make certain changes to a structure, as this teacher discovered:

Wow, when you put the plank like this, the car doesn't even get going fast enough to go up and over the bridge. Let's try making it almost straight up and down, like ski jumpers. Yikes, the car fell off halfway down; it couldn't even stay on the track! That doesn't happen to ski jumpers. I wonder what the difference is? Let's see what happens when we make the plank a little less straight.

Try out the toy area. Explore the Magna-Tiles and learn what you and your coteacher can do with them by putting together different combinations. Problem-solve when you run out of the squares. Pull out the Cuisenaire rods and investigate the possibilities that they present. Then there are the inch cubes (also known as one-inch blocks), Unifix cubes, and many other materials to rediscover or discover for the first time. And don't forget the Duplos and Legos!

I remember Tinkertoys and Lincoln Logs but none of this stuff, do you? Oh, you used the Cuisenaire rods for the "new math?" You might want to try using them with play dough, because I've seen children playing with them together. I think we should try the marble tower too. I've watched the children do it and it looks like it may be pretty difficult to build it so that the marbles don't get stuck.

Wander into the art area and experiment making different colors. How many "squirts" (from the paint pump) of yellow do you — or your coteacher — have to add to blue to get the color green you like?

I love painting like this! I never went to "nursery school," and in elementary school we worked on projects and didn't have much time to experiment with the materials. Do you want to try making different "recipes" for green? Let's see what happens if we add two squirts of yellow to the blue. I never thought of mixing paints as doing math!

Meaningful Math in Preschool

Explore the learning environment — including the outdoors — with the children and make math discoveries together.

Explore the tubs full of pine cones, acorns, pebbles, feathers, letters, pipe cleaners, and other materials. See if you can find patterns in some of the natural materials. Go to the shelves stocked with paper and familiarize yourselves with the different sizes, colors, textures, and weights.

There is some cool stuff on these shelves — acorn tops, pine needles, leaves — that's a neat idea! I wonder if we have enough materials for everyone coming to the family potluck tonight. We can invite the families to try out some of the collage materials to make a picture before we have dinner.

How long has it been since you played in the house area? Plan a party; set the table; and be sure that there is a plate, cup, fork, and spoon for each person. Look at the cookbook and get out the measuring spoons and cups and the ingredients you will need. Perhaps some babies need attention; they have been bathed but need to be diapered and dressed. Can you figure out how to get one of the tiny diapers to fit a baby and then how to fit the baby's arms and legs into the sleeper?

We could use some bigger diapers to fit the bigger dolls. I wonder if children have trouble putting the baby dolls in these sleepers? The dolls don't move the way real babies do.

Go see what's in the sand and water table today. There's rice, buckets, measuring cups, the mill wheel, and some tubes with funnels on top of them.

I've always wanted to try out the mill wheel. I've noticed that the amount of rice that comes out depends on how fast the children spin the wheel — that might be dangerous with the rice. Do you think the beans are easier to use with the tubes and funnels than the rice?

Now venture into the outdoor environment. It's hard to decide what to play with.

I'm going to start with some of the balls and see which I can bounce the highest. They didn't have such a variety of balls when I was little. Oh, you are going over to the swings? Did you ever learn to pump? I like the way it makes you go faster and higher, but it makes me seasick!

How to Use This Book

This book was the result of many conversations and classroom observations whose goal was to find ways to support teachers in providing a classroom environment that is rich with opportunities for children to explore, discover, and expand on various mathematics concepts. In this first chapter, we offered a review of the latest research in the field of early childhood mathematics; raised the issue of "math anxiety," which affects a significant proportion of both preservice and veteran teachers; and included some initial thoughts on how early childhood educators might rectify this problem.

Chapter 2 includes an overview of the HighScope approach to early mathematics and HighScope's key developmental indicators (KDIs), a brief look at the Numbers Plus Preschool Mathematics Curriculum, and a review of the different parts of the daily routine. Chapter 3 is a "materials-rich" guide to setting up an environment that invites active learning opportunities in math. In chapter 4, we explore adult-child interaction strategies, including "talking math" with children in ways that scaffold their learning.

Chapters 5–11 look at different parts of the daily routine and follow a similar format, which include What Children and Adults Do During (Planning Time, Work Time, etc.); Over There! Catch the Math During (Planning Time, Work Time, etc.); and Math in Action. These sections are described in more detail below.

What Children and Adults Do During...

This section describes what adults and children might be doing during a particular part of the daily routine (planning time, work time, etc.). We "set the stage" by drawing a visual picture of the overall scene.

Over There! Catch the Math During...

In this section, we demonstrate that math occurs throughout the day, using anecdotes that represent different HighScope math KDIs. Each anecdote is preceded by the KDI (or KDIs) observed, with the majority of the anecdotes representing the three math curriculum focal points of number sense and operations (KDIs 31–33), geometry (KDIs 34–35), and measurement (KDIs 36–37) and some examples of patterns and data analysis (KDIs 38–39). See chapter 2 for a more detailed explanation of HighScope's KDIs.

Math in Action

In this section of the chapter, we use longer scenarios to look in greater detail at what math experiences might be occurring during the children's activities, how the teacher supports those experiences, and what KDI or KDIs are observed. Here we include the KDIs in parentheticals within the scenario so you can see how children's specific language and actions relate to the different math KDIs; for example, Dom says, *"Oh, about a thousand"* (KDI 31).

Support strategies follow each scenario and provide ideas that teachers can adapt to their classrooms. These strategies encourage math learning in large-group settings; allow teachers to scaffold math learning with one or two children (e.g., small-group time or as part of a play situation during work time); and finally, enable adults to feel secure in creating an environment rich in math and "math talk."

Throughout this book, you'll find support strategies that you can adapt to your classroom so that you can support math learning and "math talk" with all the children in your classroom.

2

The HighScope Approach to Math

Allowing a young child to explore mathematics on her own terms encourages risk-taking, confidence, and creativity in mathematics.
— Gross (2006, p. 48)

Think back a decade or more ago when research pinpointed the critical importance of the early years in establishing competence in later literacy. During this time, researchers noted that children were not getting the appropriate learning experiences in preschool to support their literacy development. Early childhood professionals began implementing changes by consciously incorporating literacy-focused small-group activities. While focused, small-group activities were a critical component in the toolkit that early childhood educators assembled to take on this challenge, other tools, ideas, and strategies were also valuable parts of this kit. For example, teachers were encouraged to turn their classrooms into print-rich environments, place writing and drawing materials throughout the classroom, provide ample opportunities for children to write, support conversation among children and adults, offer their complete attention as children described things in their own way, and so much more. Following the example that was set in our approach to language and literacy, now, in addition to intentionally providing math-focused small-group activities, early childhood professionals should consider the best way to equip their own math toolkit, beginning by providing children with mathematics-rich environments (see chapter 3). Preschool teachers can also use HighScope's key developmental indicators as well as the daily routine to help frame children's math activities.

Key Developmental Indicators (KDIs)

In each of HighScope's nine curriculum content areas, HighScope identifies key developmental indicators (KDIs) that are the building blocks of young children's thinking and reasoning. The KDIs "provide a framework for supporting children's *real* activities....For adults, the KDIs give meaning to what children are doing" (Epstein & Hohmann, 2012, p. 397). HighScope's Mathematics content area comprises nine KDIs:

- KDI 31. Number words and symbols
- KDI 32. Counting
- KDI 33. Part-whole relationships
- KDI 34. Shapes
- KDI 35. Spatial awareness
- KDI 36. Measuring
- KDI 37. Unit
- KDI 38. Patterns
- KDI 39. Data analysis

See the sidebar on pages 16–17 for a more detailed description of these math KDIs and appendix A for more information about the HighScope Curriculum content areas and other KDIs.

In an effort to establish a coherent and uniform set of priorities for math standards, NCTM recommended that "all Pre-K and K work concentrate on the three topics reflected in *Curriculum Focal Points:* number and operations, geometry, and measurement," with number and operations being the primary goal (Fuson, Clements, & Beckmann, 2010, p. 1). Patterns and data analysis are referred to as *related connections*. NCTM notes that "work on patterns and data can be woven into work on the three focal points, but not with the same time investment" (p. 1).

The Numbers Plus Preschool Mathematics Curriculum, developed by HighScope (2009; see the sidebar on p. 18 for more information), is based on NCTM's curriculum focal points and connections as are HighScope's nine KDIs in mathematics. The alignment between the math KDIs and NCTM's curriculum focal points and connections can be found in the sidebar on the facing page.

Traditionally, preschool teachers have spent too little time on mathematics, typically focusing on counting, patterns, and shapes. It is universally accepted by scholars in the field of early childhood mathematics that it is "important to concentrate on number and operations and on geometry and measurement in the early childhood period, with a greater portion of time spent on number and operations" (NRC, 2009, p. 124). Number really provides the foundation for all of later mathematics. It is important to note that supporting young children's understanding of *number* is much more than just counting. As you can see in the sidebar on the right, there are three KDIs that fall under the number and operations content area: KDI 31. Number words and symbols, KDI 32. Counting, and KDI 33. Part-whole relationships. In turn, geometry (KDI 34. Shapes and KDI 35. Spatial awareness) and measurement (KDI 36. Measuring and KDI 37. Unit) support children's growing understanding of number concepts and, subsequently, each take on more significant roles in later mathematics.

Math Throughout the Daily Routine

As brain research of the last 30 years has revealed, repetition is a key factor in children's learning: "Just as hearing a story repeatedly is interesting to young children, so too young children are interested in repeatedly reciting the list of number words, repeatedly counting collections of objects, and repeatedly putting shapes together to make new shapes" (Fuson, Clements, & Beckmann, 2010, p. 5). Thus, children benefit from small-group activities that focus on mathematical concepts as well as regular reinforcement of those concepts during other parts of the day. Adults help to further cement children's understanding of basic math concepts by providing children with a variety of materials in different contexts that give them the time and opportunity to grow their ideas, develop understanding, and increase their math fluency.

Alignment of NCTM's Curriculum Focal Points and Connections With HighScope's Math KDIs

Curriculum focal points	Math KDIs
Number and operations	KDI 31. Numbers words and symbols
	KDI 32. Counting
	KDI 33. Part-whole relationships
Geometry	KDI 34. Shapes
	KDI 35. Spatial awareness
Measurement	KDI 36. Measuring
	KDI 37. Unit
Connections to focal points	
Algebra	KDI 38. Patterns
Data analysis	KDI 39. Data analysis

Throughout this book, you will become familiar with HighScope's nine math KDIs as you discover them on the message board at greeting time, in the art area, during work time, in small- and large-group times, and even during cleanup time. Let's take a short trip through a typical daily routine and see what mathematics opportunities turn up. The following examples are based on the HighScope daily routine (described briefly in appendix B). However, you can apply them to any developmentally based early childhood setting that has comparable components throughout the program day.

Greeting time

Many preschools gather together first thing in the morning to greet each other, see who is there and who is not, talk about what they might do that day, and anticipate upcoming events (e.g., field trip, holiday break). In HighScope classrooms (and in many other preschool classrooms), this is the time when children and teachers gather around to read the message board, which supports children's developing language and math skills in ways that are meaningful to them (Gainsley, 2008). For example, children read numerals on the message board; they may also count the number of links on a paper chain until the last day of school or the number of candles drawn on a cake for someone's birthday. (See the sidebar on p. 19 for a pictorial example of the message board.)

Planning and recall times

Throughout the early childhood years, the ability to plan and reflect develops gradually and with practice. When children make plans, follow through on them, and recall what they did, they are learning to express their intentions and reflect on their actions. In addition, children discover that they are capable thinkers, decision makers, and problem solvers (Epstein & Hohmann, 2012). Research confirms that when "children plan, carry out, and review their own learning activities, their behavior is more purposeful and they perform better on language and other intellectual measures" (Epstein, 2003, p. 3). In fact,

Key Developmental Indicators (KDIs) in Mathematics

KDI 31. Number words and symbols: Children recognize and use number words and symbols.

Description: Children recognize and name numerals in their environment. They understand that cardinal numbers (e.g., one, two, three) refer to quantity and that ordinal numbers (e.g., first, second, last) refer to the order of things. They write numerals.

While children are learning about number words and symbols, they are developing number sense; that is, they understand that numbers represent the quantity of real things, are manipulatable, and pertain to everyday life situations.

KDI 32. Counting: Children count things.

Description: Children count with one-to-one correspondence (e.g., touch an object and say a number). They understand that the last number counted tells *how many*. Children compare and order quantities (e.g., more, fewer/less, same). They understand the concepts of *adding to* and *taking away*.

Children develop an understanding of counting much earlier than was previously believed. Toddlers can "eyeball" (subitize) and recognize the number of items up to three, well before they begin counting. Gradually, children extend their sense of counting by developing one-to-one correspondence (naming one and only one number for each thing counted) and learning that the last number they say when counting items represents the total number of those items.

KDI 33. Part-whole relationships: Children combine and separate quantities of objects.

Description: Children compose and decompose quantities. They use parts to make up the whole set (e.g., combine two blocks and three blocks to make a set of five blocks). They also divide the whole set into parts (e.g., separate five blocks into one block and four blocks).

Children learn the idea of *part-whole* when they are exploring and manipulating materials, recognizing that a whole number (greater than one) can be made up of parts. Understanding that numbers can be composed and decomposed establishes the foundation for learning addition and subtraction, and later, division and multiplication.

KDI 34. Shapes: Children identify, name, and describe shapes.

Description: Children recognize, compare, and sort two- and three-dimensional shapes (e.g., triangle, rectangle, circle; cone, cube, sphere). They understand what makes a shape a shape (e.g., all triangles have three sides and three points). Children transform (change) shapes by putting things together and taking them apart.

While young children have an intuitive ability to recognize and match shapes before they know the shape names, they do not recognize the difference between one shape and another. Likewise, they are instinctively able to discern the difference between two- and three-dimensional shapes.

KDI 35. Spatial awareness: Children recognize spatial relationships among people and objects.

Description: Children use position, direction, and distance words to describe actions and the location of objects in their environment. They solve simple spatial problems in play (e.g., building with blocks, doing puzzles, wrapping objects).

Key Developmental Indicators (KDIs) in Mathematics (cont.)

As part of their growing spatial awareness, children first develop spatial orientation. *Spatial orientation* is understanding where you are and how to get around, and recognizing the position of objects with respect to your body. *Spatial visualization* is the ability to imagine and move objects in your mind. This may begin with a child simply remembering the location of something he or she was playing with earlier. Preschool children display tremendous development in spatial awareness.

KDI 36. Measuring: Children measure to describe, compare, and order things.

Description: Children use measurement terms to describe attributes (i.e., length, volume, weight, temperature, and time). They compare quantities (e.g., same, different; bigger, smaller; more, less; heavier, lighter) and order them (e.g., shortest, medium, longest). They estimate relative quantities (e.g., whether something has more or less).

Children have a powerful interest in comparing things. They develop their understanding about measurement as they begin to recognize that the items that they compare must have measurable properties, such as height, length, volume, or age.

KDI 37. Unit: Children understand and use the concept of unit.

Description: Children understand that a unit is a standard of (unvarying) quantity. They measure using unconventional (e.g., block) and conventional (e.g., ruler) measuring tools. They use correct measuring procedures (e.g., begin at the baseline and measure without gaps or overlaps).

Children must comprehend how measurement works before they can begin to grasp the concept of unit. By the end of preschool, most children begin to understand the idea that measuring means repeating equal-sized units.

KDI 38. Patterns: Children identify, describe, copy, complete, and create patterns.

Description: Children lay the foundation for algebra by working with simple alternating patterns (e.g., ABABAB) and progressing to more complex patterns (e.g., AABAABAAB, ABCABCABC). They recognize repeating sequences (e.g., the daily routine, movement patterns) and begin to identify and describe increasing and decreasing patterns (e.g., height grows as age increases).

Children's earliest experiences with patterns actually begin from birth as they become accustomed to the routines that their bodies establish and the schedules that they and their families settle upon. When children reach the point where they are able to begin recognizing patterns, they will first spot the repeating part (the core unit) of the pattern. This then allows them to predict what comes next in the pattern, fill it in, and extend it.

KDI 39. Data analysis: Children use information about quantity to draw conclusions, make decisions, and solve problems.

Description: Children collect, organize, and compare information based on measurable attributes. They represent data in simple ways (e.g., tally marks, stacks of blocks, pictures, lists, charts, graphs). They interpret and apply information in their work and play (e.g., how many cups are needed if two children are absent).

Young children begin using data analysis when they have a particular type of question they want to answer. The process of coming up with the answer usually involves number or sometimes measurement.

Numbers Plus® Preschool Mathematics Curriculum

The Numbers Plus Preschool Mathematics Curriculum (HighScope Educational Research Foundation, 2009) is a comprehensive set of detailed plans — activity cards — for small- and large-group activities in which adults systematically engage children in working with materials, pursuing their own investigations, and drawing conclusions.

Each of the 120 activity cards includes the following:

- **Overview.** This includes a brief description of the activity, the part of the day (small- or large-group time) it occurs, content area, topic, and more.
- **Materials.** A list of the materials needed for each child and the adult, shared materials (these would be materials for the group that every child has easy access to), and backup materials (if necessary).
- **Beginning.** Recommended words and demonstrations for introducing the activity (teachers are encouraged to adapt these to their particular group of children).
- **Middle.** A list of general ideas about what to look for in children's words and actions. To help teachers scaffold children's learning, each activity includes a developmental range chart on the back of the card with examples of what children may say or do at different developmental levels (*earlier*, *middle*, and *later*) and the strategies teachers can use to support the children at each of the levels.
- **End.** Suggestions for how to bring the activity to an end.
- **Ideas for follow-up.** Specific ideas for extending the learning to other times during the daily routine and in the different classroom interest areas.

---

making predictions (planning) and assessing outcomes (recalling) help lay the foundation for mathematical and scientific thinking.

The *planning* process is more than children simply making a choice about what they want to do; it involves "deciding on actions and predicting interactions, recognizing problems and proposing solutions, and anticipating consequences and reactions" (Epstein, 2003, p. 2). Providing a wide variety of planning strategies and activities for children ensures that the process is enjoyable rather than tedious and that you meet the different developmental levels and interests of the children in your group. Many of these strategies incorporate math concepts such as number, spatial relations, and measurement.

Here are some examples how you can incorporate such math strategies during planning time:

- Give each child a number. Roll a large die, and ask the child whose number appears to plan. (Note that this is intended as a turn-taking strategy, not a math lesson.)

- Put the symbols and names of the different interest areas on the sections of a round piece of cardboard and have three clothespins with the numerals 1, 2, and 3 on them. Ask the children to put the clothespins on where they will play first, second, and third, and then ask them to explain what they will do in each area.

- Draw a very simple map of the classroom that has the interest areas highlighted with their symbols and names. Give the children a small toy car (or dog, person, truck, etc.) to drive to the area they want to play.

The *recall* (or reflection) process is not a rote recitation of completed activities but rather a time for children reflect on, talk with others about, and display what they did during work time. As their ability to recall develops, so too does their capacity to recount past events and to think in the abstract. The strategies and activities for recall are similar to those for planning because they serve a similar purpose; that is, maintaining the interest of small groups of children while one of them is recalling (or planning). Just as with planning time, strategies for recall may incorporate different math concepts. For example, children can indicate on a simple chart the area(s) in which they played, or put something they played with in a bag and then give clues (e.g., "It's a square and it's bigger than the square Legos").

Work time

Work time is the segment of the daily routine when children carry out their plans, changing them as needed to follow their interests and solve problems encountered in play. Work time in a HighScope program is comparable to choice time or free play in other early childhood settings. Work time is both purposeful and playful. Throughout work time, children apply the concentration and seriousness of work as well as the enjoyment and spontaneous creativity of play.

Work time promotes children's natural need to explore, experiment, invent, construct, and pretend — to play! It provides opportunities for children to solve problems with materials, as in the following examples:

Math and the Message Board

Here are some math-focused examples from a message board and how preschoolers typically "read" them.

"Two writing visitors today."

"Four kids aren't here today."

"That's message 2. You can tell because of the big 2 there."

Kyra needs more bowls for her party.

▲

Henry can't get any glue out of the glue bottle.

▲

Avery, Olivia, and Tomas need more of the big hollow blocks to complete the stage they are building.

By dealing with these and other unexpected hurdles, children develop the ability to solve problems, which serves them not only later in school but also throughout their lives.

During work time, children construct knowledge when they interact, build, pretend, represent, figure out, look at books, write notes, blend colors, use magnets, set the table for a party, mold clay, weigh shells, beat drums, race cars, feed babies, and so on. When children construct knowledge, they explore and integrate learning in every content area of mental, emotional, social, physical, and creative development.

The mathematical concepts that children discover, explore, repeat, and experiment with can be observed during work time in every interest area as these anecdotes illustrate:

Josslyn counts the movie tickets using one number word for each ticket.

▲

PJ makes a sign for his store that says "Cars 5, Trucks 5, Dolls 8, Food 7."

▲

Betsy runs out of square Magna-Tiles and finds two triangles that fit together to make a square.

▲

Nygel needs a roof for his barn. He tries using one of the baby blankets. When he sees it's too small, he tries the shower curtain, which works because it is bigger.

▲

Irvin and Fatima are playing with the pegs and pegboards. Irvin says, "Mine is taller."

▲

At snacktime, Emily says, "I want all the pretzels, but none of the raisins."

Large-group time

Large-group time is a part of the daily routine when children are engaged in active learning experiences and receive adult support for their initiatives and involvement. Children manipulate materials and their own bodies and talk about their ideas and observations. They do not receive whole-group instruction in curriculum content or practice specific prescribed skills. Instead, large-group-time activities include singing, making music, moving and dancing, storytelling, story reenacting, and more. Any of these experiences may offer opportunities for children to hear and use math vocabulary, particularly words related to spatial relations, measuring, and number. For example, as children move around an obstacle course, you might hear something like "I'm getting down low to crawl under there," "Turn it upside down and let's walk on top of it," or "Go faster! I'm going to fall off!"

Similarly, while singing "Ten in the Bed," children may respond to different parts of the song with comments such as "Two dinosaurs fell out," "Six monkeys are left," or "When Jamal rolls out, there'll be no monkeys left in the bed."

Other math-related language during large-group time might include:

- *"My hands are waving over my head!"*

At planning time, the children roll a die, and whatever number they roll, they tell the teacher that many details about their plan for work time.

- *"The music's not fast; it's now slower."*
- *"I landed on a triangle this time — that means I turn around. Micky's standing on one foot because he's on a circle."*
- *"It took me five long steps to go across the circle. I wonder how many small steps it will take."*

Small-group time

Small-group times are based on adult-initiated activities in which children engage in experiences that build on their natural curiosity and interests (and, at the same time, scaffold KDIs or Individualized Educational Program [IEP] goals, or focus on assessment items). During small-group times, children make choices about how to play with materials; use language to communicate their ideas; solve problems encountered with the materials; and are encouraged by adults who participate on their level, share in the excitement of their discoveries, and scaffold (support and extend) their learning. Of course, the mathematical concepts that children learn about during a small-group time depend on the teacher's plan and the nature of the materials he or she has chosen. The following are just a few examples of what might hear children say during different small-group times.

While creating aluminum foil sculptures:

- *"I ripped mine in half."*
- *"Kacey, here's a triangle package for you!"*
- *"I scrunched it and made a ball."*

While making play dough pizza:

- *"I made the pizza pieces into squares."*
- *"Jakob, mine is the hugest pizza ever. Can you help me measure it?"*
- *"Oh, I don't want rectangles on my pizza!"*
- *"Only yellow triangles and red circles."*

When finding numbers in the newspaper:

- *"I got a big red 4! That's my number!"*
- *"This one has 1, 2, 3 all in a row."*
- *"This basketball player has a 2 and a 7 on his shirt."*

It is also crucial that all children have "experiences in which mathematics is the primary goal" and "sustained and frequent times in which they themselves enact the core mathematical content [number, geometry, and measurement] and talk about what they are doing and why they are doing it" (NRC, 2009, pp. 2, 125).

Outside time

Outside time is the part of the daily routine when children have an opportunity to use their large muscles (and loud voices), play collaboratively, try different types of equipment and materials, and explore nature. You might hear some of the following comments at outside time:

- *"We're the band. Follow me because I'm the leader."*
- *"The bird on the birdfeeder today had two green feathers in its tail."*
- *"I'm first on this swing, and then you can be next."*
- *"You need pebbles! Two handfuls!"*
- *"Okay, when I wave the flag, you start running."*

These comments from children, recorded by teachers during outside time, may appear at first to be nothing more than the typical language you hear when children feel the freedom of being outdoors. If you listen more closely, you will recognize the remarkable variety of ideas that children are constructing and processing. Learning does not stop when children go through the door to the outside.

The outdoors is not *just* a place for children to release pent-up energy or dig endless holes; it is a complete learning environment — an open-air annex to the classroom where all of the same concepts are explored. Among other things, children continue to explore mathematical concepts as they count acorns, negotiate who will be first (and second) on the swings, survey the cars in the parking lot to determine make and/or model, discover shapes in unexpected places, and speculate which of the balls will reach the bottom of the hill first. During outside time, adults do not simply supervise children for their safety; they also actively join children in their outdoor play, just like they do during their indoor time with children (Marshall, 2007).

Meal- and snacktimes

The emphasis during meal- and snacktimes is on social interaction. It is important for adults to sit down and eat with the children because it provides time for the children and adults to share casual conversation, rather than teach specific academic skills (although it is a time when teaching and learning activities occur quite naturally). Here are some examples that illustrate this point:

> *When it is Sascha's turn to pass out the plates for lunch, he is involved in one-to-one correspondence of matching plates to the people who will be eating.*

▲

> *As soon as Alana pours her scoop of snack mix on her napkin, she begins sorting it by ingredient. When it is all sorted, she counts each pile to see which has the most.*

▲

> *Curtis looks over at Allie's glass of milk and comments, "You have more milk than me; yours goes all the way up to the line."*

for children's personal creations. Keep realistic expectations of children's level of involvement and skill while supporting their learning. You'll be surprised at the abilities and math concepts that children learn during cleanup time:

> As children…put away objects in their assigned spaces, they are noticing relationships such as length, thickness, position, and quantity. For example, they see that objects are…ordered by size when they put the small, medium, and large pots away on a shelf marked with small, medium, and large outlines indicating where the pots go. They are considering concepts of number, when they notice there are three kinds of paintbrushes, each with its own container, and when they see that a marker corresponds to one hole on the marker holder. (Evans, 2007, p. 86)

▲

Now that we have addressed how essential it is for preschool children to have experiences in which math is the primary focus, the remainder of this book will help you recognize the remarkable opportunities for math learning that occur throughout the day.

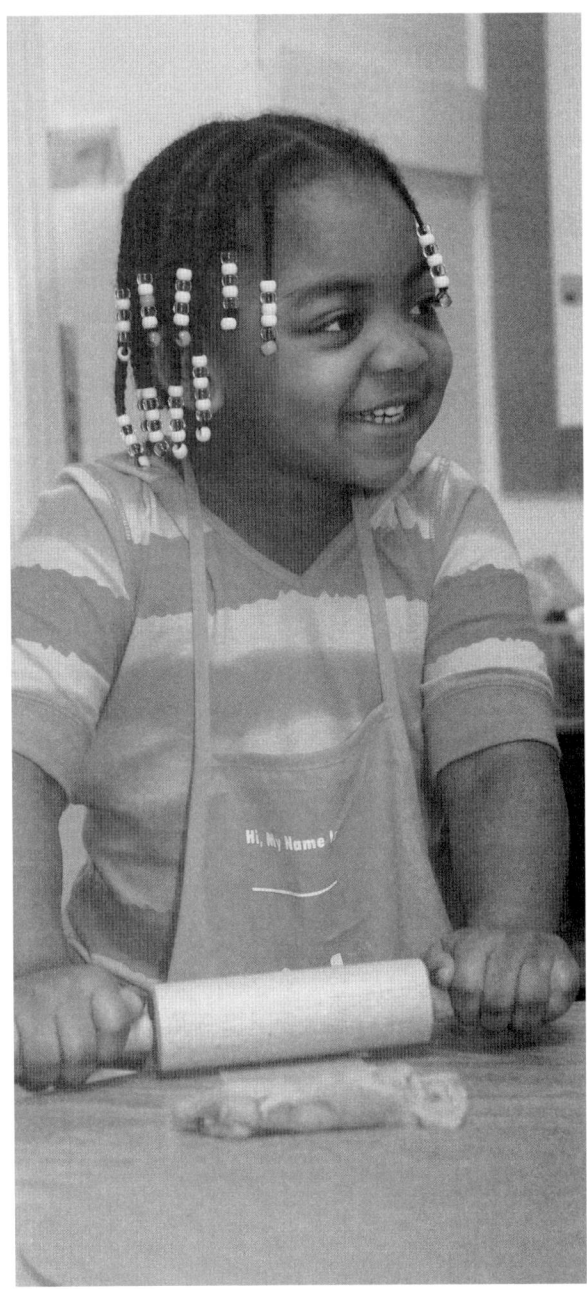

"I'm rolling out the pizza dough so it will be thin and round," says this child as she rolls out her play dough pizza during small-group time.

Cleanup time

Cleanup time should be considered a part of the daily routine and planned for accordingly (Evans, 2007). During cleanup time, children and adults work together to return materials to their storage spaces and, when appropriate, put away or find display space

3

Creating an Environment That Invites Math Learning

Most learning is not the result of instruction. It is rather the result of unhampered participation in a meaningful setting.
—Illich (1971, p. 29)

When people design a workspace for a new office, put in a new amusement park, or build a new factory, they take into account the purpose of the space — what activities will occur there, what equipment and machinery will be used, and so on. Of equal concern is the mental and physical well-being of the people who will be working or playing in that space.

The same considerations that go into designing a work environment also apply to an early learning environment. Learning, exploring, experimenting, and solving problems with materials require space and a sense of security and control. When that space is designed and equipped with math learning in mind, the adults and children who use that space every day will make mathematical discoveries as they explore the meaning of concepts such as:

- *"Three more."*
- *"Get the smallest."*
- *"It goes under there."*
- *"Way older than 30."*

Designing a Math-Friendly Environment

With the increased attention being given to early mathematics, early childhood educators are rethinking the kinds of settings and experiences they need to provide for young children to most effectively support learning in this important content area. Programs typically have a few math materials — wooden or plastic numerals for children to hold, a "number" book, a kitchen timer — because teachers tend to think of math materials in traditional terms (e.g., numerals or shapes) rather than considering how different types of materials can be used in mathematical ways (e.g., items children might count, collections whose parts can be combined in various ways).

As you think about expanding the math-related materials in your classroom, broaden your list to include anything that helps children to "mathematize" their thinking. Virtually anything that lends itself to exploring and comparing quantity, space,

Expand the math-related materials in your classroom by thinking outside the box, such as by adding an old tire to your outdoor environment.

size, pattern, and organizing information can be considered a math material, especially when adults intentionally encourage and support children as they use these materials in math-oriented ways.

The learning environment in HighScope programs, as in many early childhood settings, is arranged according to interest areas (sometimes called centers). The main areas are the block area, house area, toy area, art area, and a sand and water area. Most programs have a book area (often combined with the "quiet" area). Where space permits, programs may also include a woodworking area, movement and music area, or computer area. As mentioned in chapter 2, the outdoors is also a unique environment for learning.

The five ingredients of active learning (see the sidebar on the facing page) form the heart of the HighScope Curriculum and are incorporated into every component of the HighScope wheel of learning (see appendix C), including the learning environment. However, as you arrange and equip your learning environment, you may be particularly aware of the first three ingredients of active learning — materials, manipulation, and choice. You will find that the last two ingredients of active learning — child language and thought and adult scaffolding — come up more often in the next chapter, where we discuss the ways adults can support children's math learning.

The remainder of this chapter is organized by the main interest areas of a preschool learning environment, with suggestions of the materials you can include in these areas to stimulate or support children's math learning. Keep in mind, however, that children often move materials from one area to another and use them in ways we may not anticipate.

Block Area

Why is the block area such an obvious space for mathematical exploration? When children begin building with the blocks (with or without pretend-play materials), they are actually beginning to experiment with mathematical concepts such as geometry, measurement, number, symmetry, patterns, and more as they pile, stack, and balance blocks; turn and slide them in different ways; line them up, dump them, and carry and fit them back on the shelves; and fit blocks onto shapes outlines on the shelf and explore which one fits where.

Ingredients of Active Learning

Materials

Active learning programs offer young children an abundant supply of diverse, age-appropriate materials. Taken as a whole, these materials appeal to all the senses and are "open ended." They lend themselves to being used in many ways and help to expand children's experiences and stimulate their thought. Because preschoolers primarily deal with the world in concrete terms, materials help them pose and answer math questions, for example, sorting and counting collections to determine which group has more.

Manipulation

Active participatory learning means children handle, examine, combine, and transform materials and ideas. They make discoveries through direct hands-on and "minds-on" contact with these resources. Preschoolers are beginning to develop mental representations of objects and actions (forming images in their minds). Yet they still need to handle materials to experience their properties and see how their actions bring about change (e.g., discovering that two triangles can be rotated and combined to make a rectangle).

Choice

Children who are active learners make choices from among the materials provided by adults. They change and build on their play ideas, and plan and carry out activities according to their interests and needs and the challenges presented by teachers and parents. In an active learning setting, adults encourage children to define and solve math problems of interest to them. Exercising choice means making distinctions — what one *will (or will not)* do, choosing materials *because* they help one accomplish one's goals, deciding to use *some* or *all* the objects, and so on. Understanding and using such words and concepts are the underpinning of logic and reasoning.

Child language and thought

Children in active learning classrooms are encouraged to describe what they are doing and articulate their understanding of what they observe. They communicate verbally and nonverbally as they think about (reflect on) their actions and modify their thinking to take new learning into account. Talking helps children shape and clarify their thinking. When adults ask children to reflect and make predictions, they invite children to make sense of their world. As children talk to their classmates, they also challenge one another's mathematical thinking, and learn important concepts in the process of resolving discrepancies between their words and actions and those of their peers.

Adult scaffolding

Scaffolding means adults support children's current level of development and offer gentle extensions as she or he moves to the next developmental stage. Teachers meet this objective by acting as partners in the child's discovery process. Young children frequently solve problems using a sequential guess-try-check strategy. Adult scaffolding helps them anticipate outcomes and evaluate results. They gather and analyze data, then interpret the findings to draw conclusions about the world of mathematics. This "hypothesis testing" is the scientific process at work! Without adult involvement, the final step in problem-solving — figuring out why something does (or does not) work — does not often happen spontaneously in young children.

Source: Adapted from Epstein & Gainsley (2010, pp. 3–4).

For those children who have not had previous experience with blocks, the block area is an inviting space. Block beginners get hands-on data about math concepts such as *more, less, few, many*, and *heavy*. As they develop and become more accustomed to the shape, feel, and weight of the different types of blocks available to them, they begin to explore the world of construction and, in the process, experiment with space. They eventually confront mathematical problems presented by spatial relationships. When children begin to build stages, castles, buses, fire engines, ships, houses, drive-through restaurants, movie theaters, hospitals, forts, roads, cars, trucks, airplanes, rocket ships, bowling alleys, and more, they are beginning to explore concepts associated with balance and symmetry. Children also will add materials from other interest areas such as people figures, animals, baby dolls, food, small vehicles, blankets, or pillows to provide the additional details they need to complete their building plan (Epstein & Hohmann, 2012).

"Block areas invite mathematical exploration" is surely something you've heard someone say before and, noted in the previous paragraph, is evident. However, children do not go to the block area with math exploration on their mind. Does Jamar look at the blocks while he is planning and see geometric objects with mathematical names such as *equilateral triangle* or *cylinder*? When Kayla and Rex construct their train tracks so that they go *under* the table and then back *over* the tracks, are they thinking about spatial awareness? Probably not. Although the block area holds all the raw materials children need to explore, pretend, learn, and discover, it is up to attentive adults to scaffold (support and occasionally gently extend) children's math-related activities and discoveries so that they can internalize the math concepts they learned during previous small-group times, at home, or in the course of their play.

The following scenario exemplifies the kind of math learning that occurs in the block area. Note that throughout this scenario (and in the other anecdotes and scenarios in this book), you will see text in parentheses referring to the different math key developmental indicators (KDIs) the children's learning corresponds to. Seeing examples of the math KDIs occurring throughout the program day and in all the classroom interest areas will help develop your own awareness of the extent of mathematics learning that takes place in your preschool program. (See pp. 16–17 for a complete description of the math KDIs.)

During work time in the block area, Xavier is looking for some of the big, rectangular unit blocks to finish enclosing his doghouse. The only blocks he can find are the cylinders. He turns to Becky (the teacher) and says, "Becky, I need more of the big ones to finish my house. There are only these dumb round ones left" (KDI 34). *Becky kneels down, looks at the blocks Xavier is referring to, and says, "You mean the cylinders?" "Yeah," says Xavier unhappily, "those ones." Becky thinks for a minute, looks around, spots something, and says to Xavier with a smile, "I have an idea that might give you another way to think about your problem. Let's take a closer look at this picture." (The picture shows Roman ruins with collapsed columns, indicating that they were constructed by putting one cylindrical piece of stone on top of another.) Xavier looks at the picture and comments, "That's a wreck; it's falling down!* (KDI 35). *There are huge rocks everywhere"* (KDI 36). *Becky asks Xavier if he notices anything different about some of the rocks. "Yeah," he says, "that one is sort of round and those are round too and are standing on top of each other"* (KDIs 34, 35). *Becky waits as Xavier studies the picture a moment longer. Suddenly his face brightens as he says, "I know, I can get all those round ones left on the shelf"* (KDI 34). *"The cylinders?" asks Becky. "Yeah, yeah," he nodded, "and put them on top of each other and then finish my doghouse"* (KDI 35). *Becky stands up and says, "It sounds like you have it all figured out."*

Later that day, when Becky and Shannon (her coteacher) were planning for the following day, Becky described the exchange she had with Xavier and told Shannon that she was thinking about doing a small-group time that focused on three-dimensional shapes and their names. Shannon agreed with this idea, noting that she had noticed several other children in Becky's group playing with the castle blocks. Since this was the children's first introduction to the names for these three-dimensional shapes, Becky decided to use the basic unit blocks the children were familiar with (cones, cylinders, cubes, and spheres) with some animals or Lego people. After introducing the materials at the beginning of small-group time, Becky went around to each child and talked with him or her about what they were doing with their materials. She used the terms for the shapes in her conversations (e.g., sphere, cube, cylinder, cone) and listened to see if any children were familiar with these words or used them in conversation.

In the days following Becky's small-group time, she and Shannon provided support to children who were trying out the terms they learned during small-group time or using unusually shaped blocks in place of a typical block. For example, Macy brought the muffin pan over to the block area so she could use the small cylinders to make chocolate muffins for Shannon. She picked one out of the muffin tin; held it up to Shannon;

By supporting and occasionally offering a gentle extension, adults help children internalize the math concepts they discover during their block play.

and asked, "This is a cylinder?" Shannon smiled and said, "Why do you think it's a cylinder?" Macy responded, "Because it's round at the top and the bottom!" The teachers also made sure they were available to help the children remember the new words they had learned to describe the blocks (e.g., "Liander, you put the cone at the top of all those *cylinders*!").

Block Area Materials

Unit blocks

Unit blocks are made in standard, mathematically related shapes and sizes. In addition to the idea of *unit* (for which these blocks are named), the angles and curves that characterize some of these blocks contribute to children's earliest ideas about shape and proportion. The stability and standardized size of unit blocks are particularly appealing to early builders who link them to form roads, fences, rivers, and so forth, as demonstrated in the following anecdotes:

> *Willem makes a structure with the unit blocks exactly like the one he found in the book,* Block Builder. *He can't find the square block he needed, so he uses two triangles instead (KDI 34).*

▲

> *During home visits, the teachers took pictures of each child's house and then laminated each photo onto one of the rectangular unit blocks. When the children include them in their play and structures, the teachers hear math talk like this:*

> "The firehouse is across the street from my house" (KDI 35).

> "See, I live next to Shamar" (KDI 35).

> "This is my house; if you go over the bridge there's a park" (KDI 35).

Big hollow blocks

Big hollow blocks, which are designed to be mathematically modular, allow for the type of large-scale construction that is so attractive to young children, as illustrated by Calandra, Janie, and Kyra in the following anecdote:

> *Calandra, Janie, and Kyra plan to build a house using the "big blocks." First they get all the long ones and stand them on end. When they run out of those, they get the shorter ones and put them on top of each other to be the same height as the long blocks (KDI 36). Kyra says, "I want to go inside our house, but we need a top first (KDIs 31, 35). Go get the pink sheet — that'll work."*

Other blocks

Cardboard "brick" blocks

Cardboard "brick" blocks are made from corrugated cardboard and, if they are of good quality, are remarkably durable. The attraction of these blocks is that they allow children, particularly novice builders, to build very large structures. While the cardboard blocks lack the weight and stability of the wooden blocks, a toppling tower of these blocks will not result in the same kind of bumps and bruises that could be caused from a similar accident with wooden blocks. Cardboard blocks generally come in three sizes: small (yellow rectangles), medium (blue squares), and large (red rectangles). Let's watch now how Bonnie and Frankie use these blocks:

> *Bonnie and Frankie are playing with the cardboard brick blocks. Their plan is to build a barn for their horses, and they are placing the large red bricks on the floor to form a square "foundation" (KDI 34). As they continue to add each layer, Bonnie says, "Uh-oh, it's getting taller…how are the horses going to get in (KDIs 35, 36)?*

Creating an Environment That Invites Math Learning

At the HighScope Demonstration Preschool, the teachers print out pictures of local landmarks and then tape them (with clear packing tape) onto a variety of boxes to create the class's own set of homemade blocks.

We forgot a door! They'll have to jump really, really high" (KDI 36). *"I don't know if horses jump that high"* (KDI 36), *says Frankie.*

"I have an idea," Bonnie says. "Let's pretend they're flying horses." "Yeah," nods Frankie, "and they will fly down to their beds through a door in the roof" (KDI 35). *"Okay, but we still have to figure out how to make a door in the roof without the whole thing crashing," Bonnie reminds him. "Let's get Ms. Kay to help us. Remember she told us she fixed her roof?"*

Blocks made from milk cartons, shoeboxes, and other boxes

Blocks that you make yourself, from materials from your own home or office, can add variety to your program's block area with very little cost. You can create your own set of unit blocks with empty and cleaned milk cartons, since they are also available in standard, mathematically related sizes (e.g., pint, quart, half gallon). You can also use other boxes ranging in size from cardboard jewelry boxes, recycled banker boxes from the office, or your neighbors' moving boxes.

When using homemade blocks, consider the following to make them more durable and usable:

- Fill half-gallon and shoebox-sized boxes with materials that provide additional stability (e.g., beanbags, sand zipped in a plastic bag, crumpled paper).

- Apply contact paper (smoothing down the contact paper as much as possible when adhering) to boxes and cartons that have been tape closed.

- Take photos of some neighborhood landmarks (e.g., stores, parks, schools, firehouses), print out the photos to fit on the front of a variety of medium-sized boxes, and laminate them directly onto the boxes.

Large packing boxes

Like a sheet laid over a table or some blocks, big packing boxes can become just about anything children desire. Even better, they offer the possibility of additional design features, such as windows and doors, and they have enough surface space — inside and out — for painting. Such modifications may give you the opportunity to introduce position and direction words and encourage a discussion about shapes. As days go by, a child may see another use for the box and turn it on a different side, unknowingly practicing transformation; that is, handling, moving, and viewing something from a different perspective.

Smaller blocks

Smaller blocks, such as inch blocks (also known as one-inch cubes), organically shaped wooden tree blocks, and a variety of architectural unit blocks, may attract a builder with more detailed plans in mind. These blocks can sometimes be found in different interest areas (such as the toy area), where children who want to use them have their own space to create, out of range of the often more boisterous play in the block area.

Other construction materials

Planks

You will often find flat, long planks in the block area. Since they are longer and lighter than the long, rectangular unit blocks, planks are handy for making a roof, an extra-long bridge, or a ramp for race cars.

Arches

These pieces are broader than smaller arch blocks that are part of unit block sets. They are exactly one-half a circle, and when put together, make a circle. They are very popular as tunnels for train and car setups as well as for enclosures for classroom pets, as illustrated here:

> *During work time, J'Shawn, Brynna, and Curtis make an enclosure with big hollow blocks* (KDI 35) *so that they can play with Valentine, the class guinea pig. The children are very excited and their voices are quite loud. Valentine finds a space between the blocks and goes inside one of the smaller hollow blocks. "Oh, she's hiding!" says Brynna. "I bet she's scared because it's noisy," says J'Shawn. "How do you think we could help her feel better, J'Shawn?" asks Emily (a teacher). "Maybe she would like a couple of little hiding places where she can go when we get loud," suggests J'Shawn. "I'll get the big tunnels," says Curtis. "They are just the right size, and she can go in and out all she wants"* (KDI 35). *"She could even climb on top"* (KDI 35), *adds another child. "I'll go get two baby blankets — one for snuggling and one to put over one of the big hollow blocks so she has another place to hide"* (KDIs 32, 35), *says Brynna.*

Cardboard tubes

Cardboard tubes come in a variety of lengths (as long as 3-4 feet) and diameters and are usually sturdy enough for a child to stand on. (To find these types of tubes, visit carpet/flooring stores or fabric stores that carry upholstery fabric and ask if they have any tubes they no longer need.) In the block area, you may find children using them as part of a structure or as long, enclosed ramps for racing anything that will roll — from toy cars to small rubber balls to balls of crumpled paper.

Sheets, blankets, bedspreads, shower curtains, and tablecloths

How many of you remember building forts, caves, tents, or castles by spreading a sheet between several chairs or over a table? Preschoolers continue to discover that a large piece of fabric (often brought from the house area) often proves to be the best choice for a roof to cover their block structures. While children investigate which size fabric is the right one for their purpose, they will not only evaluate the measurable attributes of the materials they select but also demonstrate their growing awareness of spatial relations and their ability to describe the attributes of things as illustrated as shown here:

> *The children want a roof for the house they built by putting all the big hollow blocks on end. They try a tablecloth first, but one child says, "It's too small...it doesn't reach all the way"* (KDI 36). *Next they try the bedspread, but that doesn't work either: "Umph! This is big, but it's heavy"* (KDI 36), *one of the children says. They get the teacher's help to put it over their house* (KDI 35). *They crawl into the house and then, suddenly, the walls of their house start to collapse. "Oh no! Our house is falling down"* (KDI 35), *they exclaim.*
>
> *The teacher comes over, helps them put the walls back up, and then asks, "Why do you think the walls fell down?" One child comments, "That roof kept sagging and getting in the way." "I wonder what made the roof sag?" asks the teacher. The children offer several ideas, including "The blocks weren't tall enough?"* (KDI 36) *and "Was there something on it?"* (KDI 35). *"Remember how hard it was to lift?" recalls one of the children. "It was quite heavy," the teacher nods. "Maybe the roof pushed over the house"* (KDI 35), *a child speculates. "What do the rest of you think?" asks the teacher. The responses vary: "It was awful heavy; we even needed Mrs. D. to help us!" "Yeah, it could have been too heavy"* (KDI 36). *"Well, we could try the bedspread again," the teacher suggests, "or try something different." Another child suggests, "Get something that's not so heavy"* (KDI 36). *"Yeah, let's try the sheet; it's pretty light"* (KDI 36), *another child agrees. "I'll get the sheet," volunteers one child, who goes to the basket and picks one out.*

Block accessories

The following materials may be stored in other interest areas of your classroom; however, children often use these pretend-play materials in their work in the block area:

- Barn and farm vehicles
- Animal figures
- People figures representing various ethnicities, ages, and abilities
- Planes, helicopters, boats, buses, and construction vehicles
- Trains and train tracks
- Cars, roads, and street signs

As they play with these materials, children will use position, direction, and distance words, which support their developing spatial awareness, as in the following examples:

- *"Look, Valentine (the guinea pig) came out of her house"* (KDI 35).
- *"I want the barn to go next to the window today"* (KDI 35).

When you include train tracks in your learning environment, be sure that children have opportunities to set up the train tracks in the arrangement of their choice, rather than having the tracks fixed permanently to a surface. When children set up the tracks as they want, they face a variety of spatial awareness challenges as they coordinate the straight and curved track pieces, intersections, bridges, ramps, and more as illustrated here:

When children can set up train tracks where (and how) they want to, they encounter a variety of spatial challenges that encourage them to think mathematically.

- *"Move your train tracks* over *by the table — they're in the way here...no, that's too far"* (KDI 35).
- *"Watch my train. It goes up* over *the bridge and then* through *the tunnel"* (KDI 35).

House Area

The house area is a space where children can act out familiar roles (e.g., mommies and daddies) as well as roles they see outside their immediate environments (e.g., veterinarian, ballet dancer, firefighter). The house area contains child-sized appliances, adult-sized utensils, and other materials for dramatic play. According to Epstein and Hohmann (2012), "children seem to prefer the 'real thing,' probably because they see adults using spoons, mixing bowls, strainers, and so forth at home and the urge to imitate adult actions is powerful" (p. 194).

As you read about the materials you find in the house area, you will recognize that it is also a gold mine for math learning experiences, such as recognizing numerals, counting, using shapes, measuring, and more. In the house area, you might see children measuring "flour" for the birthday cake; counting out craft sticks for candles; making shape cookies from play dough; dressing baby dolls; and setting the table by putting one plate, one cup, and one fork at each place. The following anecdotes illustrate just a few of the math learning experiences you'll find in the house area:

> *One day Douglas sits on the couch and retells the story "Five Little Monkeys" to himself. He holds up five fingers, says "Five," and then holds up three fingers and says, "Three"* (KDIs 31, 32).

▲

> *Lucy makes a plan to make birthday cupcakes for Lauren (the teacher). Lucy takes the muffin tin from a shelf next to the table and gets a spoon from the container on the stove. Bella comes over and asks if she can play. Lucy says, "Sure, let's put the cupcake stuff in here" (pointing to*

the muffin tins) (KDI 35). Make sure you have a cupcake in each spot, and then we can put a candle on each one (KDI 32). Maybe we should put a bunch on Lauren's cupcake since she's probably 30-19 years old" (KDIs 31, 36).

House Area Materials

In the block area, full of shapes, sizes, and spatial relations, you can almost hear some of the materials — unit blocks, big hollow blocks, train tracks — calling out their connection to math. You may be surprised to discover that although the house area doesn't boast such obvious connections, the materials within this space offer many opportunities for math experiences.

Measuring cups and spoons and other baking equipment

Obvious material for math learning, you say? For adult use, perhaps, but preschool is not the time to introduce fractions! Instead, consider measuring cups and spoons as tools for recognizing numerals and counting and for learning to recognize an increasing or decreasing series (early algebra). If you number the bottom of each measuring cup from 1 to 4 (with 1 the smallest and 4 the largest) and hang them up in order (labeling each hook with corresponding numerals so that children can match each cup to its appropriate hook), the children will be able to see and use the numerals, learn to identify the increasing/decreasing series, and count the number of cups hanging there as Seychelle does here:

At cleanup time, Seychelle finds a measuring cup in the block area. The bottom of the cup is marked with the number 4. She takes the cup back to the house area and finds the place on the wall where the other cups are stored. The cup marked 2 is still hanging. Seychelle looks at the 4, touches it, and says, "Four." Then she touches the 4 on the cup in her hand and says, "Four." She says, "Four and four. They match!" and hangs up the cup (KDI 31).

Other cooking utensils

It is important to keep a wide variety of other cooking utensils in the house area, including those that are representative of the home cultures of the children in your program. Children recognize utensils, such as spatulas, ladles, chopsticks, whisks, rolling pins, colanders, sieves, and tortilla presses, because they see the adults in their families using them when they prepare meals at home. These utensils allow children to change the shape of materials (e.g., play dough, clay), count spoonfuls, and cut shapes:

Jamie and Fallon are using play dough, cookie cutters, and a rolling pin to make cookies in the house area. Jamie compares one of his cookies to Fallon's and says, "My circle cookie is bigger, but yours is thicker" (KDIs 34, 36). Fallon answers, "If I make mine flatter with the rolling pin, it'll be as big as yours!" (KDI 36).

Lily is using the ladle to pour water into the baby's bottle (KDI 35). "Lily! Stop! It's overfloating" (KDI 35), cries Gary, who is sitting at the other end of the table.

Pots, pans, woks, tagines, and griddles

When you stock your house area with real materials, like cooking equipment and utensils, you are telling the children that you value their efforts as they try out adult roles. Pots, pans, and similar cooking equipment also afford the same mathematical opportunities as cooking utensils:

Linnea and Julia are going to cook some spaghetti. "I'll go get the 'zgetti and sauce," says Julia. "I'll get the big pot for the 'zgetti; that's what my auntie always uses when she cooks it — the biggest, hugest pot you've ever seen!" (KDI 36).

Dishes, cups, and silverware

Children understand that if you are serving people a meal, everyone needs a plate, a cup, and some silverware. Restaurants, birthday parties, and family meals are frequent events in the HighScope Demonstration Preschool's house area and often require setting the table. Full sets of dinnerware and silver are not only necessary components of the house area but also vital contributors to the math experiences that will take place there, including "setting the table," which encourages children's understanding of one-to-one correspondence (i.e., the placement of one piece of silverware and one cup with each plate).

Cookbooks and things to pretend to cook and serve

Children prefer to create their own birthday cakes, pots of soup, pancakes, and cookies, because they are representing what they see done (and perhaps help do) in their kitchens at home. For example, compare the process of preparing a ready-made toy cake to the series of steps children follow to make their own cake from scratch with found materials as ingredients. While the children may be able to stick a few plastic candles in the toy cake, they have no choice about where to put them and little or no say in how the cake is decorated. On the other hand, the children who are preparing their own cake may first decide to consult a cookbook to make sure they know how much they need of each ingredient. Then they need to gather the ingredients they need: flour (poker chips); sugar (packing peanuts); salt (a couple of stones); and milk (from the

Real cooking utensils and equipment — the kind children see every day in their kitchens at home — allows them to try out adult roles, such as making and serving a meal.

"milk container" in the refrigerator); and, of course, candles (straws, craft sticks). Next, they get the "biggest" mixing bowl, the mixer, and the measuring cups and spoons. Let's take a look at these preschool children making a cake and see all the math involved:

> *A group of children in the house area are making a birthday cake. After they mix the ingredients and the batter is ready, one of the children observes, "It's too much for one pan; we better get two"* (KDIs 32, 36). *"But there's no other round one,"* adds another child, *"so I'll get the square one"* (KDI 34). *The children put the cakes in the oven. "Set the timer for 78 minutes, then when it's done we'll put frosting and decorations on it"* (KDIs 31, 36), *one of the children says.*

Role-play materials and props

Many of the materials for pretending and role play are stored in the house area. Pretending can involve children simply dressing up or acting in a way they *think* someone might act or something might work, for example, pretending to be superheroes. Throughout the process of preparing for pretend play and the pretend play itself, children will have math experiences, as demonstrated in these anecdotes:

> *Annan is dressing a doll with Nan (a teacher). After Annan puts clothes on the doll, he holds up a onesie and says, "This goes under it; this is baby's underwear"* (KDI 35).

> *Jesse points to one of the dress-up athletic jerseys and says, "That has a number on it*

Identifying Math KDIs

Here are a few brief anecdotes that describe children's activities in the toy area. See if you can identify the math KDIs that are at work (the answers are at the end).

1. Will nests four curves of the circle puzzle from largest to smallest.

2. Taylor points to the letter W in the box of letters and says, "Look, that's an M — it's upside down."

3. While using the nuts and bolts, Gina says, "This nut has six sides; it's a hexagon."

4. Frankie is making a Magna-Tile tower. When Sara (a teacher) comes over, Frankie says, "You are taller than my tower."

5. Maureen says that another child has more shells than she has.

6. Cherise and Hayden are stacking pegs on the pegboards. Cherise looks at Hayden's stack of pegs and says, "You used only red ones in your tower. I made a pattern on my tower. See, it goes yellow, green; yellow, green; yellow, green."

7. "Hey, Carlo, can you find five more of the red, square Legos in the tub?" asks Avi. Carlo looks in the tub, picks out some red Legos, and counts them out to Avi: "Here! One, two three, four, five — five red, square Legos!"

Answer key: 1. KDI 36. Measuring; 2. KDI 35. Spatial awareness; 3. KDI 34. Shapes; 4. KDI 36. Measuring; 5. KDI 32. Counting; 6. KDI 38. Patterns; 7. KDI 32. Counting

A Few Words About Sorting

If you had to predict what children would do with one of these collections — buttons, marbles, shells, and so on — what is the first thing you would say? Sort? You could be correct; children frequently classify (sort) objects according to one or more characteristics. However, classification falls under the Science and Technology KDIs:

Science and Technology

KDI 46. Classifying: Children classify materials, actions, people, and events.

Description: Children group similar things together. They identify relationships between things and the categories they belong to. Children look for new ways to organize the knowledge they already have and for ways to fit new discoveries into familiar categories.

If you continue to find it difficult to move "sorting" from your mental math column to the science column, first try to develop the habit of referring to the process as classification. When a word has always been associated with one particular subject, it is easier to break the habit if you change the word (but still retain the meaning). It might also help if you think back to the high school biology class in which you first learned about kingdoms, phyla, and classes.

— a number one" (KDI 31). *"I'm a Spartan basketball player," says Serena, as she puts on one of the basketball jerseys. She takes a foam ball over to where the "basket" hangs on the door. She puts the ball in the bucket and exclaims, "That's five!* (KDI 31). *I get five points when it goes inside the basket and three when it goes just outside the basket"* (KDIs 31, 35).

Toy Area

When they are in the toy area, children immerse themselves in puzzles, simple games, collections of things (keys, buttons, etc.), put-together and take-apart materials, and other manipulatives. (*Manipulatives* are materials that children can count, take apart and put together, sort, pattern, and much more.) Regardless of how (and where, as children often move materials from one area to another) they decide to use these materials, you will see that the toy area is a place where children engage with materials at varying levels of complexity as they build on their existing experiences.

The placement of small manipulatives (e.g., inch blocks, nuts and bolts, Cuisenaire rods) depends on teachers' individual preferences. Some teachers store these items in the block area, while others add them to the toy area. To help you decide what works best for your classroom, observe how your children use them.

Toy Area Materials

Small or unusual blocks

Blocks with unusual forms introduce children to new and unusual shapes and challenge their developing awareness of spatial relations, as Michael demonstrates:

Michael makes a plan to use the castle blocks and some of the Lego figures. He places one tall cone at each end of the row of arched blocks, then puts a carved piece with a round top between the two.

Under each arch, he places one of the Lego horses and says, "This is where the horses sleep. Those things (pointing to the arched doorways) are called arches and are special for horses (KDI 34). The pointy things on top are just for decoration — they're cones; they kind of look like birthday hats" (KDIs 34, 35).

Shape blocks

Shape blocks, which are often found in sets called *pattern blocks* or *attribute blocks,* are popular manipulatives that teach visual and spatial skills, geometric concepts, and other mathematical concepts. Let's take a look at what David creates with the shape blocks:

David makes a plan to work with the shape blocks in the toy area and make a flower. First he finds the skinny diamond shapes. He counts out 12 to put together (KDI 32) and form a 12-pointed star. Next, David finds some green equilateral triangles that fit perfectly in the notch formed by the two points of the star (KDI 34).

David continues to work on his flower and, when he is done, he says to Ella, "I made a flower with the shapes (KDI 34). I have wood diamonds and blue diamonds and green triangles and red squares!" (KDI 34). *Ella asks, "How did you get everything to fit together?" David shrugs and says, "I was messing around with those diamonds in the middle and found out that if I put their sides together in different ways they would make different patterns (KDIs 34, 35). Here, try it with these blue ones."*

Collections

As we mentioned earlier, there is practically no limit to the types of collections you can have in the toy area. You can start by thinking of things you might find outside that may be unique to your part of the world. In many places, you can create your own collections of natural materials, such as acorns, seed pods (that you're sure are safe), leaves (in tropical and semitropical areas, you can find palm fronds of all sizes and ferns varying in shape and size), stones, shells, and pine cones. Other things that make up collections that are interesting to children include keys, bottle caps, buttons, feathers, and any

Collections of materials encourage children to count, combine and separate, and make patterns.

other items that might make for interesting collections.

> *"Look, Amaia, here are the shells Ms. Robin told us about at message board," says Rollie. He continues, "Some are flat and some are sort of curled up and some are shut. Hey, I started a pattern: flat, curly, shut; fat, curly, curly...Oh, wait a minute, that's bad"* (KDI 38). *He looks at the pattern and then says to Amaia, "We need a shut one, do you have more (KDIs 32, 36)? Let's see how long we can make this go. There are not many curly shells so we may not get the pattern to go all the way to the shelf"* (KDIs 32, 35, 38).

Put-together and take-apart materials

This section of the toy area includes all the "fit-together" block/construction systems such as Legos, Duplos, Tinkertoys, Magna-Tiles, bristle blocks, and snap blocks. In addition, you will find pegs and pegboards, nuts, bolts and washers, Unifix cubes, Wedgits, scales/balances, and puzzles. Can you imagine all the math experiences that children can have with these materials?

Pretend-play materials

The pretend-play materials that are stored in the toy area consist of different animal collections in both large and small sizes; people figures; and sometimes small wooden village sets that include their own buildings, trees, and people. Frequently some of the small vehicles (e.g., cars, trucks, planes) may also find their way over to the house or block area as part of someone's work-time plan. Here are a couple of examples of the mathematical learning that can occur with pretend-play materials:

> *Javari is playing with the zoo animals with José (a teacher) and holds up a hippo and a rhino. He says, "It's the same color, but this guy has four horns and the other one has two"* (KDI 32).

> *Tomas and Alisha are using the farm animals and the Wedgits in pretend play. They place four animals inside one of the blue Wedgits and two animals in another blue Wedgit. Tomas says it is going to rain, so they put all the animals together into a bigger "barn" Wedgit with a "roof" to keep them dry. Alisha says, "Hey, they are all together. There's the whole family"* (KDI 33).

Games

Game pieces such as checkers, dominoes, and decks of cards are very versatile materials and, while their primary function is for game playing, they can be used in many other ways. Checkers and poker chips, for example, often turn up in children's pattern making. Dominoes can be the accepted currency for admission to a movie theater that children set up to show a "movie," and playing cards might double as tickets. Older children do enjoy playing simple board or card games, either those they have made up themselves or commercial ones in which they make their own rules. Not all preschool-aged children are ready and/or interested in playing games. Remember that those that do want to try some games may still need time to explore the pieces and parts that make up the game, including the game pieces, spinners, and board. Let's listen in on what some children say while playing games:

> *While playing Candy Land with Shannon (a teacher), Brian says, "I'm going to try and go over it," and moves his character over the bridge* (KDI 35).

▲

Matthew plays Hi Ho Cherry-O with Nona and David. He counts the cherries, "One, two, three, four," as he place them in his bucket (KDI 32).

▲

As they are playing with pieces of the Connect Four game, Jayla tells Jason, "I want to make it a pattern." Then she tells Joseph where to put the red and black pieces to create the pattern she wants (KDI 38).

Art Area

In early childhood settings, the art area is a space that is brimming with possibilities for children to express their reality and ideas and for creating make-believe. In the art area, children "stir, roll, cut, twist, fold, flatten, drip, blot, fit things together and take them apart, combine and transform materials, or fill up whole surfaces with color, fringes, paste, or paper scraps. Other children use art materials to make things — pictures, books, weavings, movie tickets, menus, cards, hats, robots… they especially want or need" (Epstein & Hohmann, 2012, p. 197). As children explore, create, and build two- and three-dimensional models, the processes they use demonstrate that, like the other interest areas we have discussed earlier in this chapter, the art area is also a rich source of math learning.

When children have a wide variety of materials available to choose from, you are more likely to see a range of learning experiences beyond those you would expect to see in an art area. Let's take a look at some children working with the different types of materials often found in preschool art areas (on the following pages) and see the math learning that takes place there.

If children have a variety of materials to choose from in the art area (as in the art area in this picture), you will see a wider range of learning and discoveries beyond those that you might expect.

Art Area Materials

Paper

The art area should include a wide variety of paper for children to explore. Going beyond colorful construction paper, recycled copy paper and different sizes of paper, add wallpaper samples, wrapping paper, and paper with different surfaces and compositions. Providing an assortment of papers not only encourages children to explore the properties of each type of paper but also leads them to discover how different painting and drawing materials interact with the different papers, as illustrated in the following anecdotes:

> *During work time, Myra is in the art area crumpling up paper into balls to roll through the large cardboard tube. As she points to each ball, she explains, "This one is the smallest because I used a really small piece of tinfoil and because it stays crumpled together; this one here is bigger because I made it from paper towel which is bigger than the tinfoil I used"* (KDI 36).

▲

> *Krystal helps Janey make a drawing of her mom by making different shapes. She says, "You use a circle for the head, lines for the arms and legs, and a…"* (KDI 34). *She turns to Moira (a teacher) who is standing nearby and says, "What's the name of that long circle shape again?"* (KDIs 34, 36). *Moira replies, "A long circle? Oh, do you mean an oval?" "Yeah," Krystal says to Janey, "Oval. That's what the body is"* (KDI 34).

▲

> *Sebastian is tracing around some of the big wooden letters and shapes. Mrs. W sits down, and Sebastian says, "See I have the S-E-B for Sebastian and I made an oval, a heart, a circle like the sun, and a triangle"* (KDI 34). *He holds up the triangle and says, "See, the triangle has three sides"* (KDIs 32, 34).

Painting and drawing materials

Just as it is important to have a variety of paper available for children to explore, it is equally important to provide a variety of painting and drawing materials. Tempera and watercolor paints give children very different results depending on the material they choose to apply the paint to. While crayons will continue to be popular (as shown in the anecdote below) and a key part of the art area's drawing materials, children discover different things about color when they have the opportunity to use markers, colored pencils, chalk, and oil pastels.

> *Allie is using the crayons to "draw a map." She says, "It takes me wherever I want to go: next door to Clinton's, underneath the street into the mall, and inside my house when I am ready for dinner"* (KDI 35).

Collage materials

Collage materials range from the natural (e.g., leaves, twigs, wildflowers, pine needles) to the recycled and scrap (e.g., packing materials, fabric scraps, telephone wire, bottle tops) and donated or purchased (e.g., craft sticks, feathers, felt pieces, paper plates, sequins, beads). Like so many of the things we assemble for preschool classrooms, safety, storage, and your imagination are all that limit the materials you collect to use for the two- and three-dimensional collages and models that children create to represent real and imagined objects as illustrated here:

> *Martin is putting the plastic peanuts around the edge of a cardboard circle. When he finishes, he colors one of the*

peanuts red and starts to count by putting his finger on the red one first, "One, two, three, four..." correctly counting until he reaches the peanut next to the red one, when he says, "Nine! It takes nine peanuts to go around this" (KDIs 32, 37).

Modeling and molding materials

Substances like modeling clay, play dough, Floam, Styrofoam, and beeswax are ideal materials for preschool programs because they suit a wide range of developmental levels. However, if a child wants to do more than explore the texture and softness or hardness of a material, it helps to also have available a range of modeling tools, such as rolling pins or thick dowels, cookie cutters, strong plastic knives, other "shapers" and cutters, and a variety of extruders (e.g., hamburger or tortilla press, garlic press) that children can use to mold, squeeze, or press materials into shapes. Often you will find that these materials are shared with the house area.

> *While playing with the play dough and the cookie cutters, Jacey says, "I made a diamond, I stuck two triangles together"* (KDIs 32, 34).

> *Judah is rolling pieces of clay into logs. He cuts one of the logs and says, pointing to two of them, "These are the same size"* (KDI 36).

Sand and Water Area

The sand and water area is popular with children and adults because it suits a variety of types of players (solitary, with a friend, in a group), handles a range of materials in addition to sand and water, and can be a rich source of math (and science) experiences and the language that accompanies them. Some of the activities that take place in this area include "mixing, stirring, heaping, dumping, digging, filling, emptying, pouring, patting, sifting, molding and splashing, as well as making pretend cakes, houses, roads, and lakes for floating boats" (Epstein & Hohmann, 2012, p. 208). While sand and water are among the most popular substances to use in the sand and water area, children enjoy the surprise when they see that there will be something different in this area for that day; it might be snow, dirt, beans, pea gravel, shaving cream, or sudsy water. A new substance in the sand and water area may mean different tools and utensils or other additions, such as large and small tubes, funnels, pulleys, buckets, and gutters.

Sand and Water Area Materials

Let's follow some children as they play in this area when the table is filled with water, sand, and beans.

Water table

> *Frances is washing her baby doll in the water. She says, "My baby is dirtier than those other babies, because she got a million bugs on her when she rolled in the dirt"* (KDIs 31, 35).

> *Jon is singing, "Underwater fire truck, underwater fire truck" as he puts his truck in the water. "Into the water goes the underwater fire truck"* (KDI 35), *he says.*

Sand table

> *Colby and Owen have a bucket full of the small dinosaurs that they bring over from the toy area. "We better dump them out*

and look for T. rex and the other scary ones," says Owen. "Maybe we should get Lissa to help. She knows tons about dinosaurs." "Yeah," says Colby, "we don't want to let any bad guys slip in with the good guys. I'll go get her" (KDI 35). *When Lissa and Colby return, she picks out all the "meat eaters." She explains that they are not scary or bad guys, but that they just eat meat, "like us"* (KDI 33). *She adds, "We're not scary!" The boys laugh. Lissa says, "You do need to keep the meat eaters and leaf eaters apart"* (KDI 33). *"We were going to put the meat eaters over there on the other side of the mountain," Colby says, pointing to the other sand table. "We'll keep the leaf eaters here where there are more trees," Colby adds as he points to the green craft sticks they stuck into the sand* (KDIs 32, 33, 35).

Bean (and other materials) table

Beans, feed corn, pea gravel, flour, rice, and other materials used in the sand and water area not only provide the adults in the classroom with an ideal situation for creating some new structures in the tables but also introduce children to experiencing the table and materials from new perspectives:

> *This morning the children do not need the message board to tell them that something is different in the sand and water area. Earlier this morning (before school began), the teachers added a two-level structure, made from cardboard boxes, in each table. There is a ramp made from a piece of gutter that goes from the top floor of one structure to the bottom of the other table. There also are a variety of tools for scooping the beans.*

The sand and water table can include a lot more than just sand and water. In this sand and water table, the children fill — and empty — different containers with beans.

During work time, many children go to this area to explore. "I put my shovel of beans through the window at the top and it went down the hole to the bottom" (KDI 35), one child comments. Another child says, "I dropped my whole spoon from the hole on top, but it got stuck inside; it's too small" (KDIs 35, 36). One child suggests, "Let's race our cars down the ramp and see which one is first" (KDIs 31, 35).

Outside Area

When we follow the children outside and look around, we should be seeing more than just a playground space where children can test their gross-motor skills and vocal chords. The outside area is an additional learning environment where children do many of the things they do in their inside space, only sometimes faster, louder, or higher. Children use the outside area differently from their inside space and find new opportunities to explore spatial awareness, measurement, number and more:

"Rachel! Watch me climb to the top of the tire climber!" (KDI 35) says Ian.

▲

Candace calls out, "Mr. G, it's my turn to swing now and I want five pushes — One, two, three, four, five" (KDI 32).

Outside Area Materials

When you plan an outside space, keep in mind that you will be accommodating different types of play. In addition to space to run, throw, and kick, there should also be paved areas for the vehicles with places to go and places to dig and cook.

Different types of play and a variety of surfaces will call for a diverse collection of materials as described here.

Fixed equipment

Include permanent structures for swinging, climbing, balancing, and sliding that also provide multiple places to hide in or crawl through as Malia and Henry demonstrate:

During outside time, Malia says to Henry, "Let's crawl through the tube and then climb down the ladder, you go first" (KDIs 31, 35).

Wheeled vehicles

The number and type of vehicles that you have really depends on your space. Do you have to walk to a park? Is your space very small or is it large but without a paved area? Start with some heavy-duty, well-balanced tricycles, Big Wheels, a wagon, a wheelbarrow, and safety helmets.

Sand or dirt area

Think about adding some of the following to your sand or dirt area to encourage exploration with mathematical concepts:

- Pots, pans, muffin tins, spoons, and other kitchen utensils
- Colanders and/or sieves
- Construction vehicles, cars, boats
- Buckets
- Shovels
- Access to water

Here is an example of one child's "math talk" while using the muffin tins and water in the sand area:

"Louisa, we need one more cup of water for the muffins, and then we will have six strawberry-flower muffins, one for each of us" (KDI 32), explains Corinne.

Loose parts

Loose parts play the same important role outside that open-ended materials play inside the

Examples of Loose Parts

Natural

Stones (heavy enough for children to use in construction projects, but too heavy to throw)
Stumps
Logs
Large branches
Small twigs
Sand
Gravel
Water
Leaves
Pebbles
Sunflowers
Seeds

Manufactured

Recycled car and bicycle tires (avoid steel-belted radials)
Pallets
Wooden or plastic crates (milk crates are favorites)
Buckets, tubs, laundry baskets
Plastic garden pots
Boxes
Gutters
Drain tile
PVC pipe
Wood: two-by-fours, four-by-fours, and planks of different lengths
Rope
Chain
Cardboard rolls and tubes of all sizes
Large- and medium-sized wooden reels
Plastic bottles
Landscape netting
Ice cream tubs
Fabric (light-weight)
Tarps or drop cloths
Hoops (Hula and others)
Weather-proof cushions
Bricks
Outdoor tools
Mesh (canvas or metal, with different-sized openings)
Chalk

Local or seasonal[1]

Sea shells
Kelp
Seaweed
Beach rocks
Driftwood
Hay bales
Bunches of wild grasses
Cornstalks
Tractor tires
Tractor seats
Troughs
Old street signs
Traffic cones
Construction debris (thoroughly sorted for safety)
Hubcaps
Car parts
Cattails and other wetland reeds
River and creek rocks
Logs
Spanish moss
Seed pods, acorns, pine cones of all sizes
Large ferns
Palm fronds
Recycled natural Christmas trees
Pumpkins

---

[1] When you are collecting loose parts, remember to respect the property you are on and ask permission before you walk off with an armload of palm fronds or hubcaps.

classroom. Loose parts can be moved, carried, combined, redesigned, lined up, taken apart, and put back together; they are frequently an element in play where you will detect the presence of math learning. Loose parts can be natural or manufactured, and, from there, the only limitations are safety, the environment you live in, and the children's imaginations.

> *Shana and Lacey grabbed the crate of hubcaps and took it to the sandbox. Shana pressed hers on the sand saying, "Mine's round." Lacey said, "Mine is round too silly. Turn it over and you might find a star, a diamond, or a square"* (KDI 34).

See the sidebar on the facing page for more information about loose parts.

Materials are important for math learning, but children need to use them with support from the adults in the classroom. In the next chapter, you'll learn how adults interact with children to encourage their mathematical explorations and discoveries.

Loose parts can include recycled cardboard tubes and a drain tile.

4

Interacting With Children to Support Math Learning

The more we help children to have their wonderful ideas and to feel good about themselves for having them, the more likely it is that they will some day happen upon wonderful ideas that no one else has happened upon before.

— Duckworth (2006, p. 14)

The strategies for identifying and supporting children's math learning that we will observe in the following chapters apply not only to HighScope programs but also to any classroom that uses a developmentally appropriate child-development curriculum where the five ingredients of active learning are present during all parts of the daily routine. In chapter 3, we addressed the first three ingredients — *materials, manipulation,* and *choice* — within the context of a learning environment that encourages math language and experiences. In this chapter, we look at *child language and thought* and *adult scaffolding* within a more general discussion of adult-child interaction strategies that support math learning.

Child Language and Thought

When children are free to go to any of the interest areas, each one filled with a variety of fascinating materials, they have lively conversations with each other and with adults. Adults in the classroom have the opportunity to encourage and support children's conversations, acting intentionally to build on (or scaffold) that incredible sense of wonder that children bring with them through our doors. Talking, conversation, and communication (among children and between a child and an adult) are essential components of a high-quality early childhood program and, more specifically, to math learning. The development of thought and language go hand in hand. Sharing their ideas with others helps young children express and clarify their own math thinking. It is therefore important that children feel free to describe what they are doing, explain their understanding of something, or clarify their reasoning behind a particular statement or action, as Nick and Zoe demonstrate in the following two anecdotes:

> *Nick looks up at Emily (a teacher who is 23 years old and very tall), then over at his mom (who is closer to 30 and quite a bit smaller than Emily), and says, "Emily, you're way older than my mom, right?"*

"Why do you think that, Nick?" his mom says, smiling. "Because, Mom, you're so much littler than her! Look, can't you see?" (KDI 36) says Nick. "I'm shorter than Emily, aren't I?" his mom says.

▲

Zoe is standing in front of her block ATM with a purse and her scarf. As she "inserts" her ATM card, she sticks a piece of paper in and out of the blocks and says to her two friends who are waiting for her, "I need lots of money if we are going shopping and then to Coney Island (KDI 32). I push my money card in and out and then the money comes out down here" (KDI 35).

Talking helps children to clarify their thoughts and make sense of their world. When children talk during their play, you are likely to hear some disputes and predictions about subjects such as whose Magna-Tile tower is the tallest, whose birthday cake has the most candles (craft sticks) on it, whose car is faster, or whose collage used the most leaves. By going through the process of disagreement, negotiation, and then resolution, children learn important mathematical concepts. Equally as important as children feeling at ease talking and expressing themselves is that the adults respect and honor children's reasoning and choice of words. As we know, preschool children explain their ideas not only through talking but also by building models, creating art, and re-creating stories or scenes drawn from their own experiences.

Noted early childhood mathematics researcher and curriculum developer Doug Clements (2001) reminds us that while people of any age can construct math knowledge, preschool-aged children have certain qualities that make their approach to math unique.

First, "the ideas that preschoolers construct can be quite different from those of

Conversations between children and adults — listening to what children think and then building on what they say — are an essential part of math learning.

adults. Preschool teachers must be particularly careful not to assume that children 'see' situations, problems, as adults do" (Clements, 2001, p. 272), as the following scenario illustrates:

Jenna is making pizzas with Ruth (her teacher). Jenna is rolling out the dough and Ruth is getting some pepperonis (poker chips) ready. "Okay, Ruth, I need six pep-a-ronis," says Jenna. Ruth puts five down on the table. Jenna counts them first, "One, two, three, four, five. Ruuuttthhhh, I need six!" (KDI 32). Ruth takes the five chips in her hand, covering them, and puts out one more saying, "Okay, that should do it. Now you have six!" "Ruth, there's only one there" (KDI 32), complains Jenna, "I said six." Ruth shows Jenna that she has the other five in her hand. Jenna shakes her head, and says, "Yeah, but I didn't see six. I only saw one!" (KDI 32).

In this situation, because Jenna could not *see* the six chips ("pepperonis") to count, she concluded that there were *not* six chips there.

The second unique quality of young children's thinking and reasoning is that children do not see or act on their world as if it were divided into separate subjects considered in isolation from one another (Clements, 2001). Though the educational system typically continues to separate content areas for purposes of instruction, research shows that learning is most effective when it is comprehensive and integrated. This principle applies to how young children best learn math, that is, as an integral part of their ongoing play and problem-solving. As Clements (2001) notes, "when preschoolers do mathematics, they really *do* it — acting with their whole being….Children's play and interests are the sources of their first mathematical experiences….From the motor and sing-song beginnings of pat-a-cake stem the geometric patterns of a 'fence' built from unit blocks and the gradual generalization and abstraction of patterns throughout the child's day" (p. 272).

Adult-Child Interactions

Let's explore the fifth ingredient of active learning, *adult scaffolding,* within the context of the adult-child interaction strategies that are especially effective at supporting math learning throughout the day. According to Epstein and Hohmann (2012), "adults, like children, have strengths and interests. In a supportive climate, adults' unique capacities and enthusiasms enrich and enliven their interactions with children, laying the foundations for authentic relationships that allow honest, effective teaching and learning to occur" (p. 72). There are several interaction strategies that will help you form authentic relationships with the children in your classroom.

Respond attentively to children's interests

Know that children's interests are the key to their learning. Give children your full attention, and make sure that your response corresponds to their interest level.

Give children specific feedback

Use your natural adult voice whether you are conversing, laughing, or sharing an observation with a child. For one thing, children have very keen, internal "fake-o-meters" and will spot the false, syrupy, cheerful voice in a second. Perhaps more important, familiar praise-driven responses such as "Nice job!" or "Pretty picture!" tend to end a conversation. Try to be more authentic and specific in your comments, offering open-ended comments that encourage continued conversation. For example, you might say something like "Mason and Lianna, it looks like you added something special to your cookies!" or "Marianne, you painted a red stripe around the edge of your paper." Responses like these will encourage children to think further about what they are doing and to say more about how and why they are doing it.

Wait patiently for children to finish and express their thoughts

Children often take a longer time than adults to organize and state their ideas. It is important to wait and listen until they finish expressing their thoughts. If they feel rushed to finish what they are saying, their thinking is cut short and they may even give up trying to converse. Don't presume to know what

When conversing with children, give them specific feedback. "It looks like you are making a pattern of dark pink and light pink beads," this teacher says to the child.

they are thinking and avoid "filling in" their ideas for them, because they may surprise you by saying something completely different. In other words, waiting for children to formulate and finish what they are saying gives us a window onto their ideas and thinking.

Ask authentic questions, and respond to children's questions honestly

When questions come up between people in authentic relationships, they are considered honest questions; that is, one person will ask another a question because they don't know the answer and are hoping that the other does (e.g., "Kati, do you know where the tape is? I need to fix a torn page in this book" or "Delia, will you show me how to make my feathers sparkly like yours?").

When you ask children an authentic question, you show them that you have genuine interest in their answer, whatever it might be.

Similarly, by responding honestly to children's questions, you are letting them know that you value what they are thinking about. Of course, you can be playful with children, provided you both know that this is a type of "silly" exchange you are having and it is their choice to carry on an interaction in this mode.

Strategies to Support Active Learning

Early childhood education research confirms that the language and demeanor adults adopt when they interact with children not only has an impact on their relationships with children and the socio-emotional climate in the classroom but also has significant impact on children's learning and development as well (Eisenberg, 2006; Hendrick & Weissman, 2010; NRC, 2000). Studies demonstrate that in classrooms where teachers are responsive, guiding, and nurturing, children take more initiative and are more likely to be actively involved and persistent in their work (HighScope Educational Research Foundation, 2009; Marcon, 2002).

In HighScope settings, there are four primary interaction strategies that teachers find most effective in supporting active learning:

1. Offering children comfort and contact
2. Participating in children's play
3. Conversing with children
4. Encouraging children's problem solving

While all four provide elements of support to math learning, we will focus on (3) conversing with children and (4) encouraging children's problem solving, which are particularly important because of their focus on adult-child communication and the critical role it plays in young children's math learning.

Conversing with children

When a verbal and/or nonverbal dialogue begins between adults and children, both contribute to it and there is a balance between the adult and child contributions. Achieving this balance, particularly when curriculum content is involved, is delicate and may require that adults make changes in their own styles of working with children. "In their interactions with children, adults are often used to being 'in control' and it may be difficult for them to learn to share control" (HighScope Educational Research Foundation, 2009, p. 3). In his groundbreaking book *How Children Learn* (1967/1983), John Holt addresses the negative consequences of adults taking too much control: "A child's understanding of the world is uncertain and tentative. If we question him too much or too sharply, we are more likely to weaken that understanding than strengthen it" (p. 95). This statement is as true today as it was more than 45 years ago when Holt wrote it, and the early childhood field recognizes the benefits that accrue when teachers learn that, when in doubt, to observe and wait for the child to initiate the conversation; when they do talk with children, it should be responsively and with respect.

Children regard familiar adults as inexhaustible sources of information, clarifications, solutions, and other important knowledge. As a result, when they go to an adult with a question, in search of help, or simply to pass on an observation, they expect an individual who is obviously interested in what they have to say, who will not interrupt, and who knows how to *listen* (note that the words *silent* and *listen* contain the same letters). You might say that in learning to converse effectively with children, adults are learning to become *active listeners*.

The following strategies will help you support children in give-and-take conversations throughout the day as they engage in a variety of activities, whether they identify a number (KDI 31), use the cardboard blocks to see who is taller (KDI 37), or put a mark on the chart under their choice of beverage for lunch (KDI 39).

Be available for conversation throughout the day. If children feel that you want to listen to them and that you enjoy talking with them, they will be more likely to talk to you. This is how children learn what an invaluable person you are to have around. It could be to listen to a story from the weekend's birthday party, participate in speculation about what kind of information is posted on the message board, respond to a request for help in figuring out a problem, or simply share in some conversational give-and-take about whatever thoughts are foremost at that moment.

Join children at their (physical) level for conversation. Have you ever been sitting down somewhere when a particularly tall person approached you with a question? Wasn't it an uncomfortable conversation, with your head tipped way back as you strained to make eye contact? Imagine how it feels to a child playing on the floor who calls a teacher over to talk or answer a question and the teacher remains standing during the discussion. When there is a significant difference between your eye level and the person you are talking to (whether it is an adult or a child), it is important that you sit, kneel, or crouch down so that you are at his or her level. This lets the person you are talking with (particularly in the case of children) know that he or she has your full attention and respect. It also affirms that the interaction is one of shared control, and that you and the child are equal partners in the process of teaching and learning.

Respond to children's conversational leads. When it is apparent that you listen quietly and with interest to ongoing conversations, children will likely address you directly or take the initiative to engage you in conversation. This allows you the opportunity to make comments and ask open-ended questions and, by doing so, encourage the child's thinking and use of language. (For more on adult scaffolding, see p. 59.)

Converse as a partner with children. When conversing with children, pass the conversational control back to the children at every opportunity. Not only does this let them know that you are interested in what they are saying, but it also encourages them to do most of the talking and thinking, and allows you to hear and understand their thought processes.

In their book *Emotional Intelligence 2.0*, Bradberry and Greaves (2009) observe that the art of listening is "a strategy and a skill that is losing ground in society" and give the following advice: "Listening isn't just about hearing words; it's also about listening to the tone, speed, and volume of the voice. What is being said? Anything not being said? What hidden messages exist below the surface?" (p. 160). Following are some hints that will help you with this process.

- **Stay with the topic the child introduces.** If you make comments, make sure they still allow the conversation to flow.
- **When talking with children, wait for them to respond before you take another turn in the conversation.** Practice getting comfortable with silence. Silence can create feelings of discomfort or a sense of pressure to fill the void. What we sometimes forget is that children need time to collect and formulate their thoughts, particularly when they are involved in an ongoing conversation.
- **Make comments and observations.** When you describe what you see children doing, you are letting them know that you recognize and respect their efforts. Since it is sometimes easy for our enthusiasm to get away with us, keep your comments and observations brief to avoid stifling the children's interest and desire to express themselves.
- **Ask questions responsively and sparingly.** When you do ask questions of children, make sure they are open ended, that is, that they do not have "one" correct answer. This will convey that you are genuinely interested in what a child has to say and that you expect to learn something you don't already know by asking the question. For example, if a child is building with the square unit blocks, instead of asking a self-evident question such as "How many blocks did you use?" you could say, "It looks like you have used up all the square blocks. I wonder what

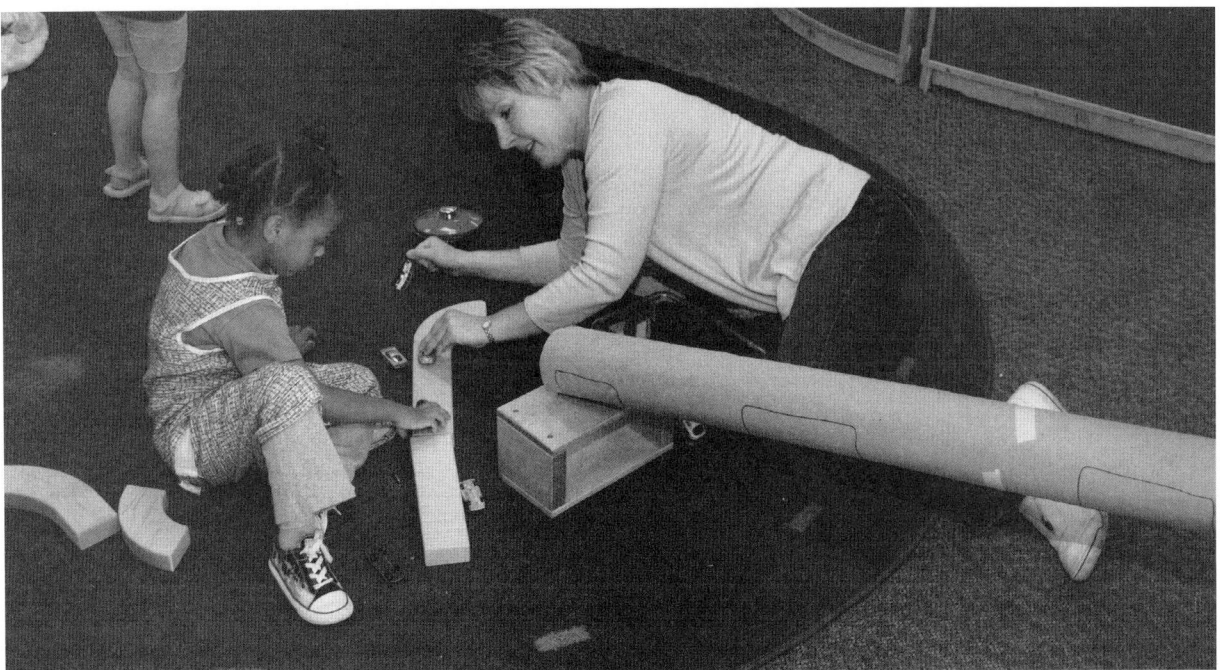

Getting down on the children's level lets children know they have your full attention and also affirms that you and the children are equal partners in the learning process.

you could use instead." This statement invites the child to explain what he or she is doing. You may be surprised to learn when the child responds, "I'm going to put two of these triangle blocks together; they make a square." Or you can use questions to explore children's thinking and reasoning. For example, if a child says one pile of beads has more than another, you might ask, "I wonder how you figured that out."

Too many questions from an adult will dampen any enthusiasm or interest a child has for trying to continue to talk with you. On the other hand, adults who know how to pose questions that are responsive to children's immediate interests, shows genuine interest in children's activities, and gives their and full attention to what children are doing and saying, have the potential for kindling rich conversations with children.

• **Relate your comments and questions directly to what the child is doing.** By first observing what children are doing, and then making comments or asking questions, you let children know you are interested in *them*, not following your own predetermined agenda. Sometimes adults will give away their lack of attention by asking a question that is totally unrelated to whatever the child is working on.

Compare how children engage with adults in the following two examples. In the first — in which the adult's comments are not related to what the child is doing — the conversation never gets started. In the second — where the teacher first observes what the child is attempting to do — math learning is taking place and the adult and child are engaging in an extended dialogue:

Gina is sticking the Cuisenaire rods into the play dough in order by height (KDI 38). *Jenny (a teacher) joins her at the table and says, "So, Gina, I hear you're moving to a new house! Are you excited?" Not only is Gina in the dark about what Jenny*

Meaningful Math in Preschool

means by "moving," her mom hasn't yet discussed the move with her yet.

▲

Geoffrey has the tape measure out and is trying to measure the large stuffed panther in the book area. He is having a hard time figuring out how to get the beginning of the tape to stay even with the panther's head. Talia (a teacher) comes over, kneels next to him, and says, "Geoffrey, it looks like you are trying to use the measuring tape." Geoffrey explains the problem he is having. Talia asks, "Would it help if I held on to the tape by the panther's head…" Before she finishes her question, Geoffrey nods his head vigorously and says, "Yeah, and I'll take this part down to the end of his tail and then look at the numbers" (KDI 37). Once they have measured the panther, Talia says, "Your dad said that you like to use his measuring tapes and rulers. Do you like to measure stuff at home?" Geoffrey explains that he is helping

To learn what children are thinking, make a comment that prompts them to describe their thought processes. For example, you might say, "I wonder how many measuring cups it will take to fill your container."

his dad build a tree house and then launches into a detailed description of all the work they did on it the previous night.

• **Ask questions or make comments about the child's thought processes.** In order to learn more about what the child is thinking, it usually works best if you comment or ask a question that prompts the child to describe his or her thought processes. For example,

- *"What makes you think that there is more water in that container than in the other?"*
- *"What do you think will happen to it if we try it that way?"*
- *"What do you think Valentine (the class pet) would do if we bought another guinea pig?"*
- *"I wonder how many red buckets of water it will take to fill the outside water table."*

One of the qualities that makes young children both a mystery and a wonder is that their brains are in high gear throughout the day — during each part of the daily routine and in every area of the classroom. Questions and statements that invite them to describe how their brains are working will stimulate their thinking even further.

Encouraging children's problem solving

Supporting children in solving child-sized problems is an important strategy that contributes to maintaining an active learning environment and encouraging young learners to work through unexpected outcomes as they explore new ideas and are faced with setbacks to their plans. Throughout the day, children encounter problems with materials ("I need more green beads for my necklaces!" [KDI 32]) and with each other ("He can't have that block — I need all the big blocks for my bridge!" [KDIs 32, 36]). When children solve their own problems, they gain confidence in their abilities while learning more about how the world works. By encouraging children to solve the problems they face with materials and to answer their own questions, you give them the confidence to try out their own solutions to other problems that they might confront or attempt something new that may have previously seemed too difficult (e.g., a puzzle).

Let's look in the art area where several children are making headbands to see how one teacher encourages a child to problem solve:

Clayton is making a "rock-and-roll" headband, and Ebelia is making a "queen" crown. They are using paper, scissors, tape, glue, glitter, sequins, markers, pipe cleaners, and more. Clayton cuts out a strip of paper, puts some silver and gold glitter dots on it, and then adds red and gold sequins. He picks out two silver and two gold pipe cleaners from the basket, tries to make some lightning bolts, but has trouble doing so. He asks Ebelia, "Belia, do you know how to make these look like lightning?" She thinks for a minute, then replies, "You mean like zigzags?" as she draws one in the air (KDI 34). *"Yeah, I can do those; give me one of the pipe cleaners and I'll show you," she says. She demonstrates the process of folding the pipe cleaner one way and then, taking one of the ends and folding it back, gives it to Clayton and says, "Do you think you can do the rest?" "Cool, yeah, thanks!" he says, and he makes the rest, attaching the pipe cleaners to the headband.*

When he tries on the headband, he realizes that it's too small (KDI 36). *He throws it on the table, calls it a "stupid, nasty*

When helping children solve problems, get down on the children's level, listen attentively, and ask open-ended questions.

thing" and then starts to cry. Karin (a teacher) comes over to him, kneels down beside him, and says, "Clayton, I can see that you are very upset about something! Is there some way I can help?" He explains about the headband, "It doesn't fit; it's too small (KDI 36). And I decorated it like a rock-and-roll guy and everything," he sniffs. Karin says, "You said it's too small. How could you tell?" He says, "Because it didn't go all the way 'round, of course!" (KDI 36). She muses, "I wonder if there is a way to make it longer?" Clayton thinks for a bit and then says, "I don't know." After a few minutes Karin asks, "Is there something we could add that will make it longer?" Clayton looks around the table, spots a strip of paper, and brings it over to Karin. "This is sort of the same size as mine, but it's a different color (KDI 36). Yeah, but if I tape it, it'll be in the back and no one will see it! Karin, will you help me make it so that it fits?" asks Clayton. "Sure," she nods. "Do we have enough tape?"

Getting down on children's level, listening attentively, commenting, asking open-ended questions, listening to conflicting viewpoints, and referring one child to another all come into play when adults support children's problem-solving process, and Karin uses many of these strategies in the brief story above. When Clayton runs into a couple of problems as he tries to put together his headband, he first seeks out another child, Ebelia, for help making lightning bolts (zigzags) from pipe cleaners, which he goes on to make. When he discovers that the headband is too small, he becomes quite upset and attracts the attention of the teacher who joins him and learns that the headband is too small. She sees that he is quite discouraged and asks him if he can think of ways to make it longer. Her question prompts him to look around, and he spots a piece of paper that will work although it is a different color. She checks with him to make sure he has enough tape, and he then solves his problem; he makes a headband that fits.

Adult Scaffolding

Adult scaffolding incorporates several of the strategies we discussed earlier; however, it proceeds *intentionally* with a teacher's considerable knowledge of how children develop and learn. For our purposes, it is best to look at adult scaffolding as a process composed of several strategies that the teacher applies at the appropriate moment. It is easier to remember, and therefore adopt, this process when you learn it as a whole. In the HighScope Curriculum, adults *scaffold* children's learning when they support children's current developmental level and provide gentle extensions as children move to the next developmental level.

Does the word *scaffolding* make you think of pipes, wooden planks, or other structures that "provide or support with a raised framework or platform" (*American Heritage Dictionary of the English Language*, 2011)? If so, you are likely not alone! It may help you to associate the word with children if you think of that building framework as simply a unique and sturdy (supportive) climbing structure designed to facilitate (extend) children's movement from one part of their playground to another, more challenging part.

So what's the difference between scaffolding children's learning and supporting it? In many ways it is the same thing and the words can often be used interchangeably. Like providing support, scaffolding includes introducing new materials and modeling new ways to use familiar materials; however, scaffolding children's learning also includes purposefully "encouraging children to *describe their actions* and *explain their reasoning,* with thought-provoking comments such as 'I wonder what would happen if...'" (Marshall, 2009, p. 1). You might do this by saying, for example, "Harry, it looks like you have picked out all frogs," and Harry might respond by describing what he did by saying "Only the green ones." Similarly, when a child makes a statement such as "My car will get to the bottom first," and the teacher responds by saying "You think your car will get to the bottom before Nick's and Nora's?" the child is likely to explain his or her reasoning: "Yeah, because it's a racing car and racing cars go faster than regular cars." It is the *intentional* use of the last two strategies (in italics above) that separates scaffolding from support.

It is important to remember that while supporting children does *not* preclude using these two strategies and, in fact, often includes it, when adults scaffold children's learning, they will also have the teacher's toolbox full of information and strategies that will be suited to a particular time of day, interest area, or individual child or group of children. This might seem like a lot to remember while interacting with children; however, you do have all of that information at your fingertips — the only difference is knowing that you will be applying it with intention. The intentionality is of critical importance in the case of math learning, now that we know that success in math learning during the preschool years predicts later success in both math and literacy. It is not always enough to just recognize an activity or behavior as mathematical; it is also important to be prepared to provide scaffolding and, if called for, a gentle extension.

▲

As demonstrated in this chapter, we not only are equipped with plenty of ideas for ensuring the presence of active learning in our classroom but also have a complete battery of strategies to support us as we scaffold all of children's learning experiences, including their mathematical ones. In the remaining chapters of this book, you will see these strategies in action as we explore math learning during the different parts of the daily routine.

Adults scaffold children's learning when they support children's current developmental level and provide gentle extensions as children move to the next developmental level.

5

Greeting Time

And then, just as Wilbur was settling down for his morning nap, he heard again the thin voice that had addressed him the night before.

"Salutations!" said the voice.

Wilbur jumped to his feet. "Salu-what?" he cried.

"Salutations!" repeated the voice.

"What are they, and where are you?" screamed Wilbur. "Please, please, tell me where you are. And what are salutations?"

"Salutations are greetings," said the voice. "When I say 'salutations,' it's just my fancy way of saying hello or good morning."

— White (1952, pp. 35–36)

Greeting time is the part of the daily routine when adults and children gather together after the children arrive, greet and talk with one another, and share news about the upcoming day. This time of day is also known as circle time, carpet time, morning meeting, and other names, depending on your program.

Just as greeting times have different names, greeting-time routines also vary, depending on how and when the children arrive, whether breakfast is served, and other factors that may be part of a particular program (Evans, 2007). As children arrive at the preschool setting, the adults who bring them may sign them in, give a message to the teacher, help them with jackets or a stubborn pair of boots, and share final hugs and kisses if they have to rush off. Time permitting, they will join their child along with other children and adults who are cozily seated on the rug reading stories, looking at picture books, drawing or writing on whiteboards, and talking with each other.

What Children and Adults Do During Greeting Time

Once all the children arrive, the first large-group activity of the day — greeting time — begins. Greeting time provides the first opportunity of the day for children and adults to gather together in a large group. In High-Scope settings, teachers and children come together during greeting time to read a daily message board. This is a social time when "children and adults collect as a community to discover what is going to happen in their classroom during their shared time together each day" (Gainsley, 2008, p. 1).

The message board is typically made up of one or more messages (each one numbered) with information for that day about the classroom, its members, and the daily routine. Teachers write the messages using a combination of symbols, simple drawings, letters, and words. Children take an active role in reading and decoding the message board, a process that supports language and literacy development as well as early math learning.

In many preschools, the main focus of greeting time is the daily calendar, which teachers use to teach children number words and symbols, counting, patterns, and other mathematical concepts (Gainsley, 2008). Children often learn these concepts through memorization and rote learning. For example, children may hear or be asked to recite the day of the week or say the numerals corresponding to the date, or a teacher may ask them to interpret small icons representing weather that are posted on the calendar. However, "research indicates that children do not understand what calendars represent until third grade or later" (Moomaw, 2011, p. 36), and looking at a daily calendar or discussing the weather is too abstract for young children; they are not likely to understand or retain the information.

Children learn best (in any content area) in an environment where they interact with real objects in meaningful situations; that is, children learn by exploring tangible materials and base their concepts about time on concrete, observable events. In the HighScope setting, the message board conveys relevant information to children about their day while simultaneously giving them opportunities to identify numerals, count, and interpret simple numerical representations such as graphs or charts (Gainsley, 2008). For example, a typical message on the board depicts the names and letter-link symbols[1] of children's classmates who are absent for the day, represented by each child's symbol with the "no" symbol (a circle with a line through it) drawn over it. Teachers might ask children how many classmates are absent (something they are naturally curious about) to incorporate counting into message board time. Similarly, by numbering each message on the message board and pointing to each number before reading the message, "teachers can also introduce the idea that numerals can represent the order or sequence of objects or events" in the context of a meaningful morning activity (Gainsley, 2008, p. 52).

Over There! Catch the Math During Greeting Time

Once your eyes and ears become attuned to the language of math, you will be amazed at how often you notice it, even as children are coming through the door in the morning:

- *"We saw two deer"* (KDI 32).
- *"We bought three pumpkins, a really big one, a medium one, and a little one"* (KDIs 32, 36).
- *"There was a bird on our window sill at breakfast"* (KDI 35).

During greeting time, there are ample opportunities to explore the three number KDIs that fall under NCTM's first focal point, number sense and operations (see chapter 2). The following examples illustrate the ways children might engage in the number

---

[1] A letter-link symbol is a picture for a word that starts with the same letter and sound as the child's name (e.g., a picture of a duck for the name Debbie).

In the first message on this message board, the teacher writes the "no" symbol over two children's names to represent that they will not be at school today.

KDIs during greeting time, providing many opportunities for adults to scaffold children's learning in these areas.

KDI 32. Counting

Damon and his grandpa come through the door. Immediately, Damon says to Tara (a teacher), "Sally had five puppies, but one died. Now there are only four."

KDI 32. Counting
KDI 33. Part-whole relationships

Lindsay (a teacher) takes off the paper covering the first message on the message board and, as she points to the numeral 1 written on the board, asks, "What do you see in the first message?" Next to the numeral 1 are the letter-link symbols representing the names of three children with ⊘ (the "no" symbol) over them. Tyler says, "Nia, Kareem, and Lily aren't here — that's three kids, one from Tara's table and two from Lindsay's table."

KDI 34. Shapes and KDI 35. Spatial awareness fall under NCTM's second focal point, geometry. Many of the books children look at during greeting time contain illustrations of both two- and three- dimensional shapes and include (and encourage children to use), direction, position, and distance words (see the sidebar on p. 65 for examples of these words). In addition, children demonstrate their spatial awareness in all the physical positioning and negotiations that take place as they decide where they want to sit, whom they want to sit next to, or when they ask someone to move because they are in the way as demonstrated in the following anecdotes.

KDI 34. Shapes
KDI 35. Spatial awareness

Antonio reads the wordless book Changes, Changes *to Lindsay (a teacher) and says, "The mommy and daddy put a red triangle block at the top." Lindsay asks, "What is it that makes the red block a triangle?" "It only has three sides!" exclaims Antonio. "Yes, triangles have three sides and they also have three points," Lindsay adds, showing Antonio the points.*

KDI 35. Spatial awareness

As message board time is about to start, Gilly says, "Sean would you move back? I can't see the message board."

KDI 36. Measurement and KDI 37. Unit fall under NCTM's third focal point, measurement. As you become more familiar with the signs that math learning is taking place, you will be able to recognize measuring experiences as easily as you recognize counting during greeting time as demonstrated below.

KDI 31. Numbers words and symbols
KDI 36. Measuring

Jody hurries through the door and announces to Tara, "I'm three; I'm big." Tara kneels down and says, "Yes, you are big, maybe bigger than yesterday."

KDI 36. Measuring
KDI 37. Unit

Tara uncovers the next message on the message board and asks, "What do you see?" Lizbet says, "The pegs and pegboards are in the toy area! We can stack them up really high. Maybe see how many pegs tall we are. We did that another day."

A joint report from NAEYC and NCTM (2010) describes KDI 38. Patterns and KDI 39. Data analysis as *connections* — topics related to the first three prekindergarten focal points (number and operations, geometry, and measurement) that guide integration of the focal points at that grade level. Once children learn to recognize patterns, they will begin to see them in the most surprising places — even at greeting time, as illustrated in these examples.

KDI 38. Patterns

Tara joins the class on the rug, singing, "It's time to put the books away and read the message board." Lucia notices the green-and-white striped tights that Tara is wearing and says, "Look! Tara has stripes on her legs; it's a stripe pattern, green-white-green-white-green-white!"

KDI 38. Patterns

Cody sits in his mom's lap fiddling with her bracelet, and says, "There's a blue bead, there's blue, there's a red one, and there are two more blue ones and another red one — Mommy, your bracelet is a pattern." "You're right," she replies.

Teachers sometimes use greeting time and the message board as a way of collecting and representing data (e.g., creating a simple chart to record children's snacktime food preferences) (Gainsley, 2008). Children learn to "read" and interpret information when it is presented in this way and may use charts, tally marks, or lists as a way of representing data they have collected in their own play and activities. The process of reading and interpreting the data usually includes some application of number (addition and/or subtraction), as demonstrated here.

Examples of Direction, Position, and Distance Words

Direction (movement)	Position (location)	Distance
up, down	in, on, under, over, on top of	near, far
move to the side	beside, between, behind	close to
turn around	in front of, in back of	far from
upside down	underneath	shortest, longest
backward, forward	middle	path
through	next to, by	
to, from	inside, outside	
away from	into, out of	
sideways	top, bottom	
across	below, above	
back and forth	same, different side	
	upside down	

KDI 31. Number words and symbols
KDI 32. Counting
KDI 39. Data analysis

The preschool class plans to walk to the ice cream store one summer day, and the manager has asked that they call in their order in advance. The last message on the day's message board is a chart with three columns. At the top of each column is a drawing of the cone and the name of an ice cream flavor. The teachers describe the flavors children can choose from — vanilla, chocolate, and strawberry — and ask them to stand up when their choice is named. One of the teachers makes a tally mark in the column that corresponds with each child's selection. "Why are we doing this? What do the marks mean?" asks Tyler. One of the teachers explains that each mark stands for a child; once they add up the marks, they can figure out what to order. Tyler says, "Well, I can tell you that only four kids want the pink flavor." "How did you figure that out, Tyler?" asks a teacher. Tyler answers, "I just looked; there are four marks under pink; that's four kids, right? You can just see that it's four. It's not like all the ones under chocolate. You can't tell how many there are because there are too many."

Math in Action: Waiting for Lindsay

This part of the chapter includes a longer scenario in which we look at what math experiences are occurring during several greeting times, what KDIs are observed, and how the teachers support those experiences (see **Support strategies** following the scenario).

In this scenario, Lindsay, one of the classroom teachers, will be out of the classroom for 10 days. What follows is a condensed account of the children's and teachers' interactions around a paper chain that Tara (another teacher) and Lindsay have made, with each link representing one of the days Lindsay will be away.

Day one

Emily (the substitute teacher for Lindsay) joins Tara and the children on the rug when it is time for the message board. Each message on the message board is covered by a large piece of paper with a numeral on it.

Tara: Do you remember that we told you that Emily would be here while Lindsay is away and she will be at Lindsay's table in the house area? Let's look at the first message (she removes the paper with the numeral 1 on it).

Tara points to the numeral 1 next to the first message, and asks the children, "Hmm…what do you see in this message?" (The message board has a drawing of a lion's face [Lindsay's letter-link symbol] with the "no" sign over it; next to it is Emily's symbol [an envelope] and then a chain made of paper links.)

Jody: Lindsay's not here, but Emily is.

Tara: (Nodding) Remember when we told you that Lindsay would be away for a while and that Emily would be one of the teachers?

Children: Oh yeah. Uh-huh. I remember.

Jody: Emily, you're sitting at my table over there (KDI 35).

Emily: That's right, Jody. Thank you for showing me where it is; I think the table is in a different place than it was when I was here before.

Tara returns to the message board, pointing at the chain.

Tara: I wonder what this chain is next to Lindsay's name.

Lizbet: It's a paper chain like we make at Christmas and put around our Christmas tree. But ours goes red-green-red-green-red-green (KDIs 35, 38).

Emily: It sounds as if you made a pattern with your red-and-green chain.

Tara stands up and holds up the long paper chain.

Tara: We made a paper chain to help us count the number of days that Lindsay will be away. The chain has one link for each school day that Lindsay will be away. Each day one of you can take off one link, and slowly you will notice that the chain gets shorter.

Gilly: It is long. Does that mean she's going to be gone a really long time (KDI 36)?

Tara: Let's count the links and see how many days Lindsay will be gone.

They all join Tara in counting as she points to each link.

Tara: Nine, 10! Whew! So how many days will Lindsay be away?

In response, she hears "10" from several children, "9" from a couple, and then some start to count again, saying the numbers to themselves (KDI 32).

Tara: Yes, she will be away for 10 school days. Tomorrow we will cut a link off the chain.

Julia: (As she points to each link and counts) So there will be only nine more days (KDI 32)!

Tara: Julia figured out that after we take one, there will be 9 left, because when you have 10 links and you take one away, only 9 are left.

Support strategies

- Tara writes a large numeral on the sheet of paper covering each message and on the message board at the beginning of each message; the numeral represents the order in which the messages will be read. When Tara or Emily refer to a particular

As the teacher and children "read" the message board, they remove the paper with the numeral on it to reveal the corresponding message. The first message on the board tells the children that Shannon (her letter-link symbol is a shamrock) will not be in today but Kay (her letter-link symbol is a kite) will be as her substitute.

message number, they point both to the numeral on the cover sheet as they remove it, and they say the ordinal number word (*first*).

- When Lizbet describes the red-and-green Christmas paper chain her family made, Emily offers Lizbet a gentle extension: "It sounds as if you made a pattern with your red-and-green chain."

- As Tara describes the construction of the paper chain, she uses the concept of one-to-one correspondence as she explains that there is one link for each school day that Lindsay would be away: "The chain has one link for each school day that Lindsay will be away."

- Tara follows up by offering a gentle extension to some of the children who are already doing some adding and taking away: "Each day one of you can take off one link, and slowly you will notice that the chain gets shorter."

- The children and adults count the links together, and Tara points to each link as they say a number.

- When they finished counting, Tara asks, "So how many days will Lindsay be away?" This gives her information about which children understand the concept of cardinality (the last number counted tells "how many"). She receives a variety of answers in response, which provides her with the information she needs. She may make a mental note to look for opportunities to do more counting with certain children or even to plan a small-group time around counting, but she does not correct or isolate any of the children who were having trouble with the concept. Once she hears what the children think, she is able to say, "Yes, she will be away for 10 school days."

- Tara explains Julia's subtraction process to the rest of the children: "Julia figured out that after we take one, there will be 9 left, because when you have 10 links and you take one away, only 9 are left." Tara's explanation validates Julia's reasoning and helps the other children understand how she was able to arrive at "9" as the number of days left.

Day five

At greeting time, Emily joins the rest of the group on the rug and sits down next to the message board. Before she can even ask the children about the message, there are children already clamoring to be the one to cut the link off the paper chain.

Emily: *There are a lot of you who think the paper chain is under the second message.*

Lily: *Yup, it was under there yesterday and on Carrie's birthday day.*

Emily: *Let's look and see what's under the first message.*

Emily pulls away the paper covering the first message (a cover sheet with the numeral 1 written on it) and points to the numeral 1 on the message board. Next to it are drawings of two houses.

Lily: *Oh, that means two home days. (She points at the drawing) One, two (KDIs 31, 32).*

Emily: *Yes, Lily, you counted two houses, which means two home days.*

Emily counts out loud "One, two" as she points to the houses again.

Emily: *After you go home today, you will have two home days and then it will be time to come back to school.*

Kareem: *I bet that the paper c is under message two (KDIs 31, 35).*

Tara: *You think it's under number two, the second message, this time?*

Kareem: *For sure...see how the paper is sticking out?*

Emily: *Let's see what's under message number two! (She removes the paper chain hiding under message two.)*

Kareem: *Yeah! There it is! I want to cut off the chain.*

Tara: *Okay, Kareem, but first let's take a look at it.*

Tara takes down the paper chain and holds it up.

Sean: *Shorter means less days (KDIs 32, 36).*

Emily: *(To Sean) Yes, it is shorter than when we started. There are fewer days left. Do you remember how many links there were when we started?*

Sean: *Yeah, we counted and there were 10 (KDI 31).*

Emily: *I wonder if anyone remembers from yesterday how many there were.*

Jody: *I counted and it was six (KDI 32).*

Several of the other children also say six (KDI 31).

Emily: *Kareem, would you cut off another link?*

After Kareem cuts off the link, Lucia jumps up.

Lucia: *I want to count; I haven't had a turn.*

Tara: *Lucia is going to count to find out how many days are left until we see Lindsay.*

Lucia begins to count, touching each of the remaining links.

Lucia: *One, two, three, four, five. There are five left — five days until Lindsay gets back (KDI 32).*

Tara: *Yes, five more days.*

Support strategies

- Emily and Tara support children's understanding of ordinal numbers by referring to the messages as the *first* and *second* messages.

- Emily acknowledges ("Let's look and see…") Lily's assumption based on what occurred on previous days — that the paper chain would be under message two: "Yup, it was under there yesterday and on Carrie's birthday day."

- By using Lily's words, Emily lets Lily know that she recognizes Lily's understanding that the drawing of the two houses means the children will be having two home days: "Yes, Lily, you counted two houses, which means two home days."

- Emily responds to Sean's statement "Shorter means less days" by affirming that the paper chain is getting smaller and that there are fewer days left until Lindsay returns. By restating the connection that Sean has made, Emily reassures him that the time is getting shorter.

Using a paper chain, this teacher helps a child count the number of days left before the other teacher comes back from vacation.

- Tara explains that Lucia is counting to figure out how many days are left on the chain and that each chain link represents one day. When Lucia completes counting, saying that the last number she counted — five — was the number of days until Lindsay would be back, Tara nods her head and simply says, "Yes, five more days."

Day nine

The children and teachers are in front of the message board and have just read the first message.

Tara: *(As she points to the numeral 2 on the message board)* Ready for the second message? Antonio, you are waving your hand; would you like to guess what the second message is about?

Antonio: It's the chain for Lindsay.

Tara: What makes you think it's the chain?

Antonio: Because it was there yesterday and the day before that and a lot of days before that.

Tara: Now I'm going to take the paper off, and we will see if it is there again.

She pulls off the paper covering the second message and points to the numeral 2 on the message board. There, next to the numeral 2, is what is left of the paper chain that they have used to mark the days of Lindsay's absence.

Tara: What do you notice about the paper chain today?

Della: Two left. Only two (KDI 32)!

Tara: *(Smiling)* Just two more days. We still have one more thing to do to this chain before we can go on to message three.

Gina: We have to take a link away. Can I do it?

Tara holds up the remaining link.

Children: *(Together)* One! Only one more day until Lindsay comes back (KDI 32)!

Antonio: *(Holding up one finger)* One day, just like one finger (KDI 32).

Tara holds up her finger next to his and points to the numeral 1 on the message board.

Tara: Just like that, Antonio, one and one.

Support strategies

- Today, instead of using the cardinal form of the number word when referring to "message number two," Tara instead uses the ordinal form when she asks the children if they are ready for the "second message." As always, she points to the numeral 2 both on the cover sheet and on the message board, so that there is no question about which message she is referring to.

- When Tara asks, "What do you notice about the paper chain today?" Della answers, "Two left! Only two!" Tara repeats Della's response, "Just two more days."

- Once Gina cuts off the next-to-last link, Tara holds up the remaining one so the children can observe and conclude for themselves that Lindsay will be back in just one day. Being able to see and reach a mathematical conclusion on their own is an important learning experience for children; it is also immensely satisfying! And so is recognizing and correcting an adult's mistake. For example, if Tara had deliberately said, while holding up the one remaining link, "So we still have two days until Lindsay gets back," some of the children would likely have said, "No! One more day. There's only one link left!" Tara could have then acknowledged, "Oops! Only one more day until Lindsay gets back."

6

Planning and Recall Times

If we give [children] the opportunity to organize their ideas in their own way, provide contexts that are meaningful to them, and allow them to communicate their ideas with their own language, then we have the opportunity to be surprised and excited by their understanding of mathematics.

— Gross (2006, p. 49)

Planning and recall times are the parts of the day that bracket work time. Together, these three segments form plan-do-review — the heart of the HighScope Preschool Curriculum. Although planning time and recall time each last only 10–15 minutes, the wide variety of planning and recall strategies available to teachers present quite a few opportunities for math learning.

What Children and Adults Do During Planning Time

In their small group with a teacher, each child plans what he or she will do during work time that day. The planning process encourages children to think through what they want to accomplish during that time. This process is governed by purpose and intent, as children actively make choices about what they will do based on their own interests, rather than choosing from options determined by the teacher. In her book, *Making the Most of Plan-Do-Review*, Nancy Vogel (2001) describes the process this way: "As children plan, they construct mental images of occurrences that have not yet happened, and they begin to realize that they can act on those images and ideas and carry them out… this helps them develop a sense of control and become independent and self-confident" (p. 15). The adult role during planning is to support, facilitate, and encourage children to communicate their plans in a variety of ways, expand on their ideas, and modify their support strategies as children become more experienced planners.

When adults encourage children to plan in different ways, they are supporting their brain development — different experiences cause them to use different thought processes. In addition, offering a variety of interesting and exciting planning games and experiences can help to maintain children's interest and enthusiasm throughout the process. While these activities keep this time appealing for children, the adult's focus is on listening to and supporting children's ideas as they learn to express their intentions. Thus,

For this planning time, the teacher asks the child to roll a die, count the number of dots on the top of the rolled die, and then tell her that many things about her plan.

it is important to keep the activities short, simple, and appropriate for the children's experience and developmental levels.

Over There! Catch the Math During Planning Time

Since planning and recall times are the shorter segments of the daily routine, the opportunity for math learning may not present itself daily. Over the course of several days, however, evidence of math learning will regularly appear as adults engage children in conversation, posing open-ended questions that prompt math language, as demonstrated in the following planning-time scenarios.

> **KDI 31. Number words and symbols**
>
> *For planning time, Jenny (the teacher) has put the different symbols for the classroom areas on carpet squares. She asks the children to jump on the carpet square that represents the area they plan to be in during work time. Allegra jumps to the carpet square with the house area symbol on it. Jenny says, "I see you landed at the house area. What do you plan to do there?" Allegra answers, "First, Lizzie and I are making banana chocolate muffins. Second, we're making tacos. And third, we're having a surprise birthday party for..." (she whispers in Jenny's ear). Jenny comments, "It sounds as if you will be busy. I'll come by later and see if you need some help getting ready."*

KDI 32. Counting

While looking through the "spy glass" at the art area during planning time, Tomoko says, "I'm going to paint." Cathy (a teacher) asks, "Do you know what paints you will use?" Tomoko answers, "I'm using two paints."

KDI 33. Part-whole relationships

When it is his turn to plan, Athi says, "Block area — animals." Cathy (a teacher) responds, "You want to play with the animals in the block area? I wonder which animals you are going to use." "All the big lions," Athi says.

KDI 34. Shapes
KDI 35. Spatial awareness

Today for planning the children put a sticky note with their name and symbol on something they are going to play with and then return to the table to give clues. Corinna goes first and her clues are "It's in the art area, next to the beeswax. I'm going to roll it flat and make circle and square cookies."

KDI 31. Number words and symbols
KDI 35. Spatial awareness

It is Lana's turn to plan and she picks up the camera, aims it at the art area, and pretends to take a picture. Cathy (the teacher) says, "Lana, it looks like you took a picture of the art area. What would you like to do there?" Lana says, "I want to paint a rainbow first. Then I'll put a house in the middle under the rainbow and some flowers beside the house."

KDI 36. Measurement
KDI 37. Unit

Asked about his plan for work time, Nick says, "Bo, me, and Rolf are going to build a tower." Jenny (a teacher) replies, "I remember that you three worked on a tower yesterday at work time." Nick says excitedly, "Today we're going to make it taller than me. We'll keep building until the blocks are higher than my head."

KDI 36. Measurement

Zanni is listing what she needs to wear to be a princess. "That is quite a long list," says Cathy (the teacher). "Do all princesses need to wear so many things?" Zanni replies, "No, but I'm the oldest and that means I get to wear the silver shoes, the longest gloves, and the crown."

KDI 38. Patterns

When Filipe joins his group at the round planning table, he says, "Look. We're sitting boy, girl, boy, girl, boy, girl all around the table."

KDI 31. Number words and symbols
KDI 38. Patterns

When it is Mayte's turn to plan, she says, "First I want to go to the art area and make a necklace for my mommy." Cecelia (a teacher) responds, "What do you think you will use to make the necklace?" Mayte answers quickly, "Oh, the blue, green, and purple beads. I'm going to make a pattern that goes blue, green, purple, blue, green, purple, like that. Those are mommy's favorite colors."

KDI 33. Part-whole relationships
KDI 39. Data analysis

Bella is the last one at her table to plan. She has been listening carefully to hear where the other children plan to play. Before she tells her teacher her plan, she says, "Sally and Tim and Leo went to the house area. And Jonah and Amani are in the block area. I'm going to the block area too, so it'll be the same in both areas."

This child uses a "spy glass" (in this case, a sparkly cardboard tube) to see what she plans to do during work time. "First I'm going to the block area and then I'm going to the woodworking area," she tells her teacher.

Math in Action: Planning With Sammy and Claire

In this scenario, we take a closer look at the math experiences during a particular planning time for Jenny's small group, the KDIs observed, and how Jenny supports these experiences.

> *It's planning time at Jenny's table. She has been encouraging some of the more experienced planners in her group to add details to their plans. Today she is using a large foam die as part of her planning strategy. (To modify the foam die, she covered the sides with five and six dots with paper and drew two and three dots to replace the five and six dots.)*
>
> *As the children settle at the planning table, Jenny holds up the die and says, "Today for planning time, we are going to take turns rolling this die. After you roll the die, we'll count the number of dots on the top. That's the number of things you can tell us about your plan."*
>
> *She hands the die to Sammy who rolls a three. Sammy glances at the die and says, "Three" (KDI 32). "You didn't even count the dots!" exclaims Jenny. "I could just tell it was three" (KDI 32), replies Sammy. "Okay, Sammy, tell us three things about your plan today," says Jenny. Sammy begins, "I'm going to play in the block area with Justin." Jenny holds up one finger. "What will you and Justin do in the block area?" Jenny asks. "Make a rocket," answers Sammy. Jenny holds up another*

74

finger and says, "Play in the block area and make a rocket — what are you going to use to build the rocket?" Sammy replies, "We're using a lot of the big blocks" (KDI 32). *Holding up three fingers, Jenny says, "Okay, I count three things: playing in the block area, building a rocket, and using the big blocks. It sounds as if you are ready to go!"*

Next it's Claire's turn; she rolls a four and immediately holds up four fingers as she says, "House area — dress-up, book area, then the block area, then the toy area" (KDI 32). *Jenny smiles and says, "Wow, you said a lot of different things, Claire! You don't have to do four different things! First, would you just tell me a bit more about what you are going to put on after you get to the house area?" Claire responds eagerly, "First, I'm getting the purple dress.*

Second, I'll put on the sparkly shoes and then I have to find the flower scarf for my hair (KDI 31). *That's three things I'm putting on"* (KDI 32). *"Oh, and if I get there in time," she adds, "I want to wear the long leopard gloves"* (KDI 36). *Jenny says, "Claire, you are putting on four things! I'll come over later and see if you need help with those gloves."*

Support strategies

- Jenny uses a counting game — rolling a die — as a planning strategy to get the children to think about multiple components of their plan. By using a die, where dots represent numbers, she is furthering the children's understanding of number (e.g., •• is the same as 2 is the same as *two*).

This child strings a bead on yarn that is attached to different classroom area signs to signify where he plans to play during work time.

- Jenny helps Sammy keep track of the number of ideas he tells her about his plan by holding up a finger for each idea.

- When Sammy finishes his plan, Jenny summarizes the things Sammy listed, noting that he had named three: "Okay, I count three things: playing in the block area, building a rocket, and using the big blocks."

- Jenny uses the ordinal word *first* to help children think about the sequence of the things they plan. In addition, she uses the word *after* to bring Claire's attention to the order in which she'll carry out her plan.

- When Claire rattles off four things she plans to do, listing four different interest areas, Jenny clarifies that she is interested in learning more about what Claire will do in one specific area. This tells Claire that Jenny is interested in the details of her plan. Claire pauses after listing three items; Jenny does not interrupt but waits until Claire says, "That's three things...." Knowing that the gloves can be difficult to put on, she assures Claire that she will be over to help her with them.

What Children and Adults Do During Recall Time

Recall time immediately follows the work-time and cleanup segments of the daily routine and brings closure to those activities. Just as planning time provides children with a chance to imagine activities that have not yet occurred, during recall, children's brains perform the following important processes: drawing on mental images, reflecting on experiences, associating plans with outcomes, and talking with one another about their discoveries and actions (Epstein & Hohmann, 2012).

As children become more practiced in recall activities, their ability to remember details of what they did increases. This process of "picturing and thinking through past events helps children learn that they have control over their actions and choices and helps them view themselves as capable doers" (Vogel, 2001, p. 132). In addition, as children become more adept at forming and talking about mental images, their consciousness expands beyond the present and concrete ("here and now") and they begin to think in the abstract ("then and there"). As children are able to provide more detail, adults can offer support by encouraging them to talk about the order (sequence) in which they accomplished their work-time activities.

As you do for planning time, be sure to vary the props and games you use for recall. Providing a variety of recall experiences causes children to draw on different parts of brain; one day they use a graph, the next day they may draw a picture or dictate their story to the teacher for a recall book. It is important to keep the recall games and experiences simple and short, as you do with planning time. Your time and attention should be focused on supporting children as they talk about their work-time experiences rather than on playing the game. When you are a keen listener during planning and recall, you are more likely to notice math language from children.

Over There! Catch the Math During Recall Time

As children recall, they not only choose what to talk about but also use their own words to describe what occurred. When children recall,

Planning and Recall Times

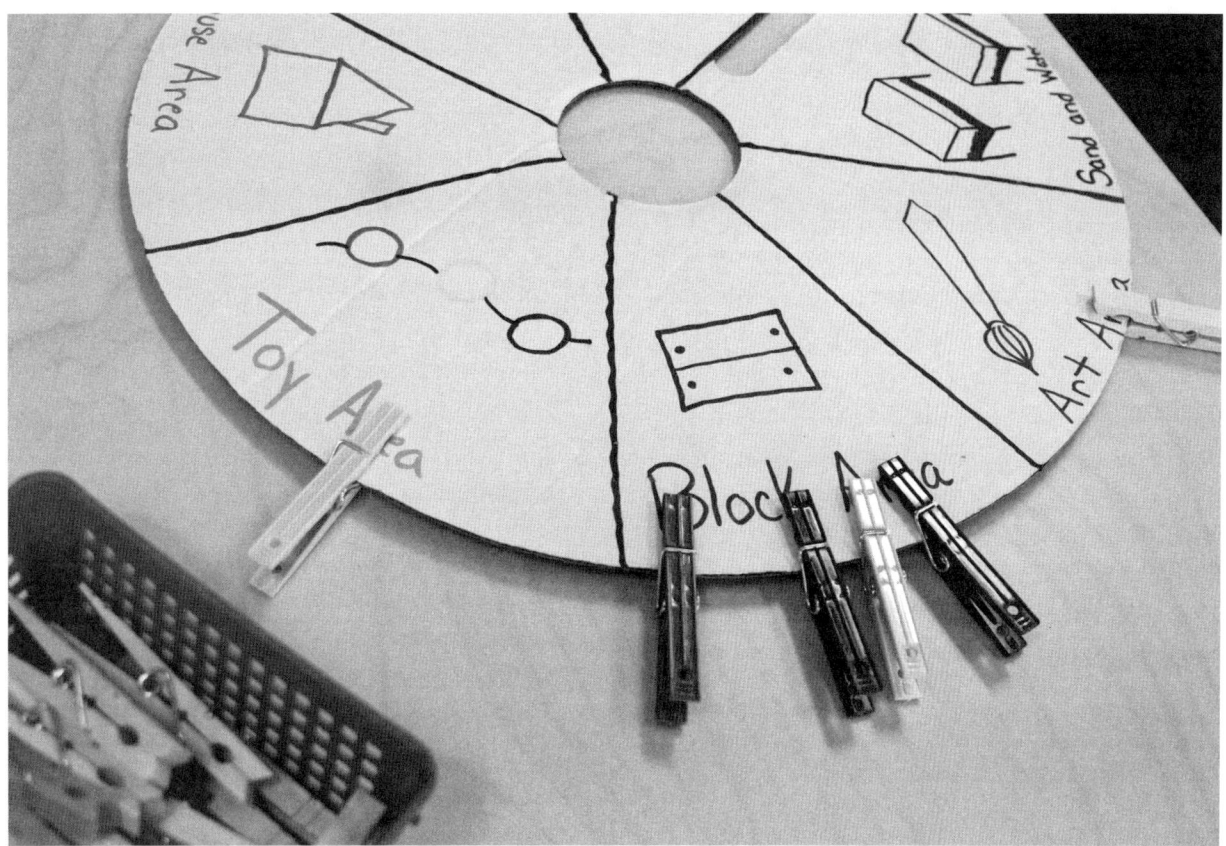

For this recall time, the teacher divides a cardboard circle into the different classroom areas and asks the children to use clothespins to mark (and then talk about) which area they played in during work time. After the children recall, the teacher uses this cardboard circle with clothespins to support the children as they analyze the data: how many children played in the different areas and which areas had the most (or fewest) children.

they discover that they can make things happen, learn new things, and solve their own problems. Recalling daily eventually results in children's increased higher-order thinking (e.g., synthesis, evaluation, etc.) as they remember more and more about what they did. Let's look at the math KDIs identified in some preschool recall times.

KDI 31. Number words and symbols

Using the clothespins with the numerals 1, 2, and 3 on them and a cardboard circle divided into the different interest areas, Allegra puts the number 1 clothespin on the block area section and says, "First, I built a stage in the block area with Tomoko, Athi, *and Claire," she says. Then she puts the number 2 clothespin on the house area section and says, "Second, I went to the house area to make chocolate-banana muffins." Finally, she puts the number 3 clothespin on the computer area section and says, "Third, I went to the computer area."*

KDI 32. Counting

When it is his turn to recall, Rolf points to the toy area. Jenny says, "You played in the toy area? I remember that you and Chung had a plan to do that. What did you bring back that you played with?" Rolf opens his bag and, while pulling one item out at a time, says, "I have three things: one — Magna-Tiles, two

— marble racer, and three — policeman. Mostly Chung and me made rockets and a space station with the Magna-Tiles."

KDI 32. Counting

Matthew says, "During work time, I was mostly in the block area and I played with three different kinds of blocks: the unit blocks, the castle blocks, and some of the tree blocks."

KDI 33. Part-whole relationships
KDI 36. Measuring

After she draws what she did at work time, Allegra says, "I used all the big horses and some of the little horses."

KDI 34. Shapes

Today the children are drawing recall pictures on individual whiteboards. Justin holds his up and says, "I drew the oval of the blue rug."

KDI 35. Spatial awareness

As recall is about to begin, Russ and Corinna are talking about who is sitting next to whom for recall time. Russ says, "I'm sitting next to you but not in the same chair." Corinna replies, "I'm sitting next to you in the same chair and Justin is in the middle."

KDI 35. Spatial awareness

Bo describes one of the things (a foam die) he played with during work time: "It's green and black and inside my hat."

KDI 36. Measuring

Cathy has put two sand timers — one large and one small — on the table for recall. She has told the children they can choose a sand timer when it is their turn to recall. Pointing to the larger one, Lizzie says, "It takes longer for the sand to go down so you can talk longer."

KDI 36. Measuring

There are three bags on the table: large, medium, and small. Cathy asks the children to think about the things they played with during work time, and when it is their turn to plan, to select the bag that is closest in size to what they played with. Nick says, "I took the smallest bag because I played with Legos."

KDI 31. Number words and symbols
KDI 32. Counting
KDI 36. Measuring
KDI 37. Unit

Lana and Zanni are quite excited and choose to recall together. Lana says, "We built a tower using the big hollow blocks." Zanni adds, "Yeah, it was so tall we had to get on a chair to add the last block!" Lana continues, "And then we connected all the links — all of them — to measure how tall it was." Zanni then says, "Lana held one link at the bottom and Cathy had to lift me up so I could hold the links up to the top of the tower. Then we counted all the links. Cathy had to help." Jenny asks, "Do you remember how many links high it was?" Lana and Zanni look at each other for a minute and Lana says, "I think it was four-six high."

KDI 39. Data analysis

Today for recall time, the children throw beanbags into buckets that were each labeled with the name of an interest area. When they are finished, Sammy looks into each bucket and exclaims, "Hey no one played in the art area, at the sand and water table, or in the book area!"

Math in Action: Using a Recall Chart

In this scenario, Cathy (a teacher) asks the children to mark with an ink stamp in the classroom areas they worked on a recall chart. Let's see all the KDIs Cathy observed in this recall time and how she supports children's experiences.

On a recall chart, this child marks with a stamp where he played and then tells his teacher two things he did there.

For recall time, Cathy makes a chart with six columns, each labeled with the name and symbol of one of the six classroom interest areas. She puts the chart in the middle of the table and says to the children, "Today, I have a recall chart and an ink stamp so you can put a stamp in each area you worked. Then we can find out how many children played in the different areas of the classroom."

Cathy says, "Blake, where did you play today?" "The block area and toy area," answers Blake. Cathy says, "Oh, tell us about what you did in the block area." After Blake shares his recall with the group, Cathy slides the chart toward Blake along with an ink stamp and ink pad. Blake puts a stamp in the columns for the block and toy areas.

Cathy turns to Rowan and says, "Rowan, where are you going to put a stamp?" Rowan stamps each column and says, "All the areas." "You stamped all six areas. Did you go to every one?" asks Cathy. After Rowan nods her head, Cathy continues, "Tell us what you did in one of the areas."

After the other children take turns recalling, Cathy holds up the chart and says, "Wow, we have lots of stamps in some of the columns. How can we tell how many children played in each area?" Several children say, "Count the stamps!" Cathy responds, "Okay. How many stamps do we have in the block area column?" Cathy points to each stamp as she and the children count together: "One, two, three, four, five. Five stamps means that five children played in the block area," she says. Cathy writes the numeral 5 at the bottom of that column and explains, "I wrote the number 5 so we can remember." After she and the children count the stamps in the remaining columns and write the numerals at the bottom, Cathy says, "Let's figure out where the most children played." "The block area" (KDI 39), calls out Rowan. "How did you figure that out?" asks Cathy. "It has five stamps. That's the most" (KDIs 32, 39), answers Rowan. "Yes," says Cathy. "The most stamps are in the block area column. Now, in which area did the smallest number of people play?" she asks. "The water table" (KDI 39), *says Blake. "Look at how many stamps are in the water table column," says Cathy. "None"* (KDI 32), *replies Rowan. "Nobody played at the water table today? That's strange. You put a stamp there, Rowan," says Cathy. "Oh, yeah. I forgot," says Rowan.*

Support strategies

- Cathy uses a chart to incorporate counting and data analysis into recall time.

- Cathy explains what the marks on the chart represent: "You can put a stamp in each area you worked in. Then we can find out how many children played in the different areas of the classroom."

- When Rowan says that she played in all the areas, Cathy quantifies what Rowan means by saying, "You stamped all six areas."

- Cathy helps children interpret the chart: "Wow, we have lots of stamps in some of the columns. How can we tell how many children played in each area?" Several children say, "Count the stamps!"

- Cathy counts with the children by pointing to each stamp as she says a number and then repeats the last number she says to indicate the total quantity.

- Cathy encourages the children to interpret the data: "Let's figure out where the *most* children played" and, later, "How did you figure that out?"

- Cathy extends children's thinking about counting and quantity by using the terms *most* and *smallest number.*

The children put sticky notes (labeled with the classroom areas) in the column with their name/letter-link symbol before they recall what they did during work time. After all of the children have recalled, they can use the chart to see who played where (data analysis).

7

Work Time

Observed as a whole, a group of 18 children at work time can exhibit what looks like a lot of random movement. Observed individually, however, each child's actions generally fit an internal logic that is related to the child's particular purpose. So while children move from place to place and talk with one another along the way, their actions more often than not serve some purpose they have set for themselves.
— Epstein & Hohmann (2012, p. 279)

Work time, the heart of the plan-do-review cycle, is the part of the daily routine when children turn their plans into actions. Throughout this part of the daily routine, you will observe children making choices, selecting materials, discovering new concepts, and finishing projects they started. During work time, children initiate active, hands-on experiences that allow them to construct their own knowledge — the kind of knowledge and learning represented by the HighScope key developmental indicators (KDIs).

What Children and Adults Do During Work Time

Work time — the "do" part of the plan-do-review process — is the longest and arguably the most important component of the High-Scope daily routine. During this time, children carry out a purposeful sequence of actions (that they have discussed earlier during planning time) and also may pursue new ideas or plans as they play. Children are free to work with any of the materials in any of the interest areas and may move materials from one area to another. Such flexibility further encourages children's natural tendency to use materials creatively and intentionally build upon activities that interest them, unconstrained by activities predetermined by adults. Children are also free to work alone, in pairs, or in small or large groups at work time. They may interact with teachers and/or with other children. These shifting patterns allow them to pursue their own mathematical inquiries (e.g., "How many cups of water can I add before it spills over the top of the bowl?") as well as to address the challenges posed by others (e.g., "I say I'm taller than Laslo and he says he's taller than me").

During work time, children apply both the concentration and seriousness of work and the enjoyment and spontaneous creativity of play. The role of adults at work time is to keep their focus exclusively on the children, rather than being distracted by classroom chores or paperwork. Adults maintain an awareness of the status of children's plans in the event a child needs help getting started or has switched to a different plan. In this way, adults remain alert to any children who may need reassurance, comfort and contact, or encouragement before moving on, continuing, or delving deeper into what they are doing.

As you will see in this chapter, work time is that part of the HighScope daily routine when the adult-child interaction strategies reviewed in chapter 4 become very clear. The strategies that an adult chooses to use with a child or group of children depends on children's developmental levels, their observations of children's individual needs and interests, and the types of play taking place in the classroom. During work time, adults

- Play as partners with children.
- Interact with children, sharing give-and-take conversation about the play.
- Scaffold children's learning.

In the first chapter, we met a pair of teachers who spent several hours exploring their classroom materials — discovering and rediscovering the possibilities that these materials had to offer. As a result of this hands-on exploration, they developed the kind of familiarity with many of the materials that one only gets after actually manipulating and playing with them. Adults' knowledge and familiarity with the materials in the classroom is an important factor in their ability to effectively scaffold children's math learning.

Teachers should also ask themselves this simple — but important question — "Am I ready to play?" Learning to play with children provides you with much of the information you rely on when determining interaction strategies as well as children's developmental levels and for informing your thinking as you plan for the next day. When you join children's play, you show children that you value and respect their interests. Before you join children at play, pause to be sure that you can join during a "natural opening" in their play and that your entrance (at the children's level) will not disrupt the flow of action. Be sure that the children retain control of the direction of the play; if you want to offer new ideas, keep them within the context of the play that is already occurring. Address children during play in a way that acknowledges and respects the roles or characters they have assumed (e.g., "Bus driver? Do you stop at my school?").

Work time is a nearly bottomless mine of examples of children's math learning and experiences. This is largely due to the presence of at least five interest areas (e.g., art area, block area, house area, sand and water area, toy area) that hold different materials and therefore stimulate a wide variety of plans, some of which will inevitably involve mathematics. In order to illustrate the range and variety of math experiences that take place throughout a typical work time, this chapter will focus on individual interest areas. As you read through the sample anecdotes for each of the interest areas, think of scaffolding strategies that might be appropriate for that situation (bear in mind that children may also look to other children for support and assistance or may prefer to work alone).

Over There! Catch the Math in the Art Area

Imagine a child's first glimpse of the art area — an enchanted space filled with sticky, sparkling, colorful, bristly, mysterious delights. (For more information on materials in the art area, see pp. 42–43.) The variety of choices

Because work time is the longest period of the daily routine, children have ample time to explore and use any of the materials in the interest areas. These two children spent their work time building a very tall tower.

is staggering, and the prospect of being free to use *any* of the materials in *any* way they choose may make many children feel as if they have entered a magical area of the classroom. (Of course, teachers must always be aware of the needs of individual children who may have tactile sensitivities and make any necessary accommodations to support them.) From this same space also come countless examples of math experiences in progress and, as a result, opportunities for you to offer support during those experiences. Let's take a look at some of things children are doing when they are in the art area, keeping the math KDIs in mind and listening for hints of "math talk" in children's comments.

KDI 31. Number words and symbols

Avery writes the numeral 6 on the picture he is drawing. "That's my 6," he says, pointing to the number on his shirt.

KDI 32. Counting
KDI 33. Part-whole relationships

Mina says, "I glued five things: One, two, three here and two over there."

KDI 34. Shapes

Ronin rolls the tape on the table and says, "This is like a wheel because it's a circle."

KDI 35. Spatial awareness

Olivia paints a border on the picture and says, "Look! I put spots all around my picture."

KDI 36. Measuring

Dom is making a headband and says, "This is not the right size."

KDI 36. Measuring
KDI 37. Unit

Henry rolls out logs of play dough and places them in two piles. "I measured them against this," he says, holding up a piece of pipe cleaner. "Those are shorter and those," he says, pointing to the other pile, "are longer."

Math in Action: Brianna's Collage

Brianna is making a collage using some of the new materials the teachers introduced this morning on the message board at greeting time. Elena (a teacher) is already in the art area helping Diego put his paper up on the easel. Let's see how Elena uses adult-interaction strategies to support Brianna's and Diego's mathematical learning during this work time.

Elena: *Is this high enough, Diego?*

Diego: *No, it's too high (KDI 36).*

Elena lowers the paper. She then goes over to Brianna, who is already collecting the materials for her collage, and crouches down beside her.

Elena: *Brianna, your plan was to make a collage. It looks as if you have all of the tubs and baskets from the shelves on the table!*

Brianna: *I didn't know how much of the leaves, pods, and other outdoor stuff I would need and I wanted to get the sequins first (KDI 31).*

Elena: *Oh, are there certain sequins you're looking for?*

Brianna: *I want a bunch of the circles because they are the biggest (KDIs 32, 36). I also want to find some diamonds and stars (KDI 34).*

Elena: *Brianna, I wonder if you can tell me about how many "a bunch" of circles is? If I know that, maybe I could help you by finding that many.*

Brianna: *(Thinking for a minute) A bunch is a lot, like this many (she cups both hands together to show it takes two hands to hold a bunch). Maybe seven or eight (KDI 32).*

Elena: *Okay, I'll look for those.*

Brianna: *Will you look for the diamonds and stars too? Make sure they look like this star, because there are different stars in there — I want this one with five points (KDIs 32, 34).*

Elena: *Sure. Do you want me to see if I can find any other shapes?*

Brianna: *Well, maybe some triangles, but no squares or rectangles. The triangles have just three sides (KDI 34). When I get all the sequins, I'm worried that if I put them on the table they might get lost.*

Elena: *Do you want to put them in something?*

Brianna: *Here (handing Elena two lids). You can put all the circle sequins on this lid and the rest of the sequins on the lid next to that tub (KDI 33). First I am going to glue this shaggy moss on the paper, and then I'm going to get three leaves (KDIs 31, 32). I'll have to hold them up against the paper before I glue them and make sure they don't cover up all*

Work Time

A child counting squirts of paint to put in a paint cup is one math experience you'll observe in the art area.

the moss (KDI 35). I need some sticks, but all that other stuff (pointing to other tubs) is too heavy for the glue stick, heavier than the moss (KDI 36).

Brianna gets some tape, picks out three sticks from a handful that she has grabbed, and then puts the rest back in the tub (KDIs 33, 35). She breaks one and then holds that one up to the others so she can make them about the same length (KDI 37).

Brianna: *Elena, can you help me put tape on my sticks? If you hold the stick, I'll put on the tape.*

Support strategies

- Using measurement and spatial awareness terms, Elena asks Diego where on the easel he wants to hang his paper: "Is this high enough, Diego?" By doing this, she provides Diego with an opening to respond mathematically, using language related to measuring: "No, it's too high," Diego responds.

- Elena describes the quantity of materials she sees that Brianna has collected on the table: "It looks as if you have all of the tubs and baskets from the shelves on the table!" Brianna's explanation to Elena ("I wanted to get the sequins first") shows Elena that Brianna is developing a sense of cardinality.

- Elena asks Brianna for additional information regarding the sequins she wanted ("Oh, are there certain sequins you're looking for?") thinking that, in response, Brianna might name the shapes rather than the colors of the sequins she wants, and she does: "I want a bunch of the circles because they are the biggest. I also want to find some diamonds and stars."

- Elena uses Brianna's words ("I want a bunch of the circles") to ask Brianna about her thought processes and to gently extend her thinking: "I wonder if you can tell me about how many 'a bunch of circles' is?" Elena explains that this will allow her to help Brianna find what she needs. Pausing after this question, Elena gives Brianna ample time to collect her thoughts before she answers.

- In not challenging Brianna's final answer or grilling her with additional questions, Elena is letting her know that she respects both her reasoning as well as the process she followed to arrive at her answer.

- When Elena extends a request from Brianna ("Will you look for the diamonds and stars too?") with a question related to what Brianna is doing ("Do you want me to see if I can find any other shapes?"), she is both continuing to show Brianna that she is interested in what she is doing *and* encouraging her to continue her math talk by possibly naming more shapes.

- Elena follows up with an open-ended question that includes some math language: "Do you want to put them *in* something?" Sometimes inserting the smallest math word in a comment or open-ended question may encourage the child to use one or more in response. Elena's use of the position word *in* as part of her question may have resulted in Brianna's use of *on* and *next* in her response: "You can put all the circle sequins on this lid and the rest of the sequins on the lid next to that tub." The increase in Brianna's use of position words (in addition to direction and distance words) indicates that she is developing spatial awareness.

Over There! Catch the Math in the Block Area

Walk into any early childhood setting, and you will likely find that the block area is a very busy place. Perhaps the blocks are becoming a city bus or a complex rail system complete with rail crossings, bridges, and tunnels. Or, a city may be taking shape around the rubber strips the children use as roads for their small vehicles (see pp. 30–34 for additional information on block area materials). In the block area, children explore, build, and pretend. Consider the potential for math learning when children explore in the block area in the following ways:

- Piling, stacking, restacking, and balancing blocks (KDIs 32, 33)
- Sliding blocks and turning blocks different ways (KDI 34)
- Loading blocks into boxes or onto trucks; lining up blocks, dumping them, and carrying and fitting them back on the shelves (KDI 35)
- Turning different pieces of railroad track different ways to make the track go in a particular direction (KDI 35)
- Figuring how many blocks are needed to make a second tower as tall as the first tower (KDIs 32, 36)
- Discovering what order to stack blocks of different sizes so they don't topple over (KDI 35)

Listen for the "math talk" and be alert for math experiences taking place as we look in on the events that occur over several days in the block area. As you read each example below, think about the support or scaffolding strategies you might use.

KDI 32. Counting

Jill (a teacher) helps Kyra get some unit blocks to build a "horse barn." Jill holds up the arch blocks and asks Kyra, "Do you need any arches?" Kyra nods and holds up three fingers as she says, "Five."

KDI 33. Part-whole relationships
KDI 35. Spatial awareness

Brianna gets a handful of train tracks out of the tub and puts them in a pile, then takes another handful out and puts them in a second pile. She puts some tracks together, turning several of them to get them to fit. Once she has enough tracks to form an oval, she puts the remaining tracks back in the tub.

Children often bring materials from other areas into the block area. The children here wear construction hats (from the house area) and use magnifying glasses (from the toy area) to take a closer look at the grain in the blocks.

KDI 34. Shapes

Sabrina cannot find a square block of a certain size to match the picture in the Jack the Builder *book. Instead, she suggests, "If I put these two triangle blocks together like this (she demonstrates), it makes the square I need."*

KDI 35. Spatial awareness
KDI 36. Measuring

Mason hides some blocks for Jill (a teacher) to find and says, "You have to look behind something taller than me."

KDI 36. Measuring

Aisha and Jamee are putting cars down the big tubes. Aisha says, "Mine went faster — all the way to the rug."

KDI 36. Measuring
KDI 37. Unit

Ronin and several other children are jumping from a line they taped on the floor and then measuring their jumps with a long, rectangular unit block. After his jump, Ronin measures and says, "I jumped three blocks!"

KDI 38. Patterns

Elena makes a green-yellow-green-yellow-green pattern with the inch blocks. Diego sits down with her and adds another yellow and green block to extend Elena's pattern.

KDI 32. Counting
KDI 39. Data analysis

Amare and Olivia are playing a game with balls, and Amare keeps score by making tally marks on a sheet of paper. He counts the marks and says, "I have four and you have two. I have more so I'm winning."

Math in Action: Razi and Henry Build a Road

Razi and Henry have plans to build a city. They look toward the block area where other children have already begun building. Jill, the teacher, supports the children in their road building experience while extending their mathematical thinking.

Razi: *There's not enough room. Those kids have to move their blocks to give us more room* (KDI 32).

Razi tells the children in the block area to move out of his way. Jill approaches Razi and asks him to explain the situation, noting to herself that Razi anticipates a spatial problem.

Jill: *You think you will need lots of room to build your road? More room than this? (Jill indicates a space available in the block area.) Look around the room. Do you see another space big enough for your road?*

Henry: *We could build it over there in the book area. It's bigger* (KDIs 35, 36).

The boys build their roads by alternating one long and one short unit block.

Henry: *Look at our road, Jill! Can you tell we made a pattern* (KDI 38)?

Jill: *It looks like you alternated one long and one short block across to the shelf.*

Razi: *We need to turn our road. We need the curved blocks* (KDI 34).

Razi selects some of the curved blocks and adds them to the road, but then he notices that the road now curves into the bookshelf. He flips the block so it curves away from the bookshelf and continues building (KDIs 34, 35). *As the boys continue to add blocks to the road, they flip and turn the blocks to avoid other obstacles in the way* (KDIs 34, 35). *Jill notices and comments on the boys' work.*

Jill: *You made a long road. You started with both short and long straight blocks but then you had to change to curved blocks so your road would turn away from the bookshelf and the couch. It can be tricky to figure out how to flip and turn the blocks to make them go in the right direction.*

Razi: *Next thing we have to do is put up some buildings* (KDI 35).

Support strategies

- Jill hears Henry and Razi express frustration because there is no space left in the block area to build the city they had planned: "There's not enough room. Those kids have to move their blocks to give us more room." When Jill joins the boys, she restates their problem — "You think you will need lots of room to build your road? More room than this?" — and gestures, urging them to look around the room for another space. Henry points out the space in the book area, noticing that it is "bigger." Jill's open-ended questions encourage the boys to do some problem solving around their spatial requirements and to think about playing in a space where they had not used blocks before.

- When Henry points out that the blocks they used for their road created a pattern, Jill confirms their statement: "It looks like you alternated one long and one short block across to the shelf."

- As the boys realize that they need curved blocks to avoid the shelf, Jill calls attention to the attributes of the blocks they have used so far: "You made a long road. You started with both short and long straight blocks but then you had to change to curved blocks so your road would turn away from the bookshelf and the couch."

- As the boys were adding the curved blocks to build their road around the obstacles in the room, Jill acknowledges the boys' solution to the spatial problem, pointing out the strategy they used: "It can be tricky to figure out how to flip and turn the blocks to make them go in the right direction." This supports the children's developing spatial awareness.

Over There! Catch the Math in the House Area

The house area offers children the chance to "try out" some of the roles, occupations, and practices that they see at home and in their community. For example, the children might decide to have a birthday party. Some of the children might look in the cookbook for a recipe to bake the cake, while others might set the table and decorate and a few other children might wrap presents (see pp. 34–38 for additional information on house area materials). At the HighScope Demonstration Preschool, the house area is often home to several busy "mommies" and their "babies" or a place where various princesses and superheroes fix themselves food. It has also served as a doctor's office, a veterinary clinic, and a restaurant — complete with a drive-through window and equipped with a credit card machine, cash register, cash (paper strips and rocks), and plenty of hungry customers. In all of these situations, you will find opportunities to support math concepts.

Let's look at the math learning — through the KDIs — children experience when they engaged in the following activities in the house area:

- Trying on clothes and shoes: KDIs 31, 35, 36
- Dressing baby dolls: KDI 35
- Making change for customers: KDI 32
- Adding ingredients: KDIs 32, 36
- Making phone calls: KDI 31
- Using kitchenware and utensils: KDIs 31, 32, 36, 37
- Setting the table: KDIs 32, 33
- Running a movie theater, restaurant, or store: KDIs 31, 32, 34, 35, 36, 37, 39
- Figuring out how much a baby weighs: KDIs 36, 37

Are your math antennae tuned in? As you follow the children in the house area in these anecdotes, consider how you might scaffold their math learning, looking for ways to support their actions and ideas, and, where appropriate, to provide a gentle extension.

KDI 31. Number words and symbols

Fatima is putting on the silver shoes. When Nikki comes up to her and says she wants to wear them, Fatima replies, "I'm first. You can be after me; you can be second."

There are many math experiences happening here during this cooking session in the house area. The children are reading a recipe, measuring and adding ingredients, and using cooking equipment.

KDI 32. Counting

Avery tells Marcella (a teacher) that he is making cherry babka like his Bubbe makes. He counts four of the seven marbles (cherries) he puts on the play dough before he rolls it up.

KDI 32. Counting
KDI 34. Shapes
KDI 35. Spatial awareness

Ronin is slicing play dough pizza and says, "I cut it in the middle." Dom says, "Ronin, I want the piece next to that one (he points) because it has my favorite: two triangle cheeses and three circle pepperonis."

KDI 31. Number words and symbols
KDI 36. Measuring

Kyra and Brianna are playing with the baby dolls. Kyra says to Brianna, "My baby is three. How old is yours?" Brianna replies, "She is four. She is older than yours; yours is younger."

KDI 31. Number words and symbols
KDI 37. Unit

Henry looks at a recipe in the cookbook. He says to Sabrina, "For cupcakes we need: sugar — nine cups, M&Ms — two scoops, milk — one glass."

Math in Action: Jamee Has a Checkup

Over the past several days some of the children have been talking about trips to the doctor for their checkups. Today Olivia and Jamee are setting up a doctor's office in the house area and they ask Elena (a teacher) to be the doctor. Let's see how Elena supports and extends Olivia's and Jamee's mathematical thinking as the "doctor."

Jamee: I'm your next patient, and I need a checkup. That means you have to listen to my heart, check my eyes, and look in my ears and mouth.

Elena goes through all the parts of the checkup and is about to send Jamee on her way when suddenly she remembers something important.

Elena: Jamee, I forgot to find out how tall you are. I should measure you to see if you've grown. Would you please go stand against the wall?

Jamee stands against the wall and places her hand on the top of her head to mark how tall she is on the wall.

Jamee: I'm this tall (KDI 36).

Elena acknowledges Jamee's gesture.

Elena: You are that high on the wall. When I was little, my mom put tape on the wall to mark how tall I was. Should I do that?

Jamee agrees, so Elena marks her height on the wall with tape. By this time other children want to be measured also. As the children stand against the wall, Elena marks their height with a piece of tape with their name written on it. When Elena has finished measuring several children, Jamee steps back and looks at the wall.

Jamee: You know, kids are lots of different sizes. Some are taller and some are shorter (KDI 36). *This is the tallest. (She points to the highest piece of tape.) It's Katie. She's the tallest* (KDI 36). Can we see how tall she is with the measuring tape (KDI 37)?

Support strategies

- Just as Jamee's checkup was wrapping up, it occurred to Elena that she could show Jamee one way to measure. Saying to Jamee that she "forgot to find out how tall she was," Elena asks Jamee to go stand against the wall.

- Elena marks the children's heights with tape so they will be able to see and compare them. Jamee notices the difference in the levels of the pieces of tape and comments, "kids are lots of different sizes. Some are taller and some are shorter."

Over There! Catch the Math in the Sand and Water Area

Whatever material, structure, or apparatus occupies the sand and water area on a particular day, you can be sure of having a front-row seat for a variety of math experiences (see pp. 43–45 for additional information on materials in the sand and water area). Unfortunately, some teachers may decide not to include this interest area in their classroom due to space limitations or the fact that it can be "messy." However, it is important to remember that young children learn with *all* of their senses, and sensory play contributes in significant ways to brain development: "Think of it as food for the brain. Stimulating the senses sends signals to children's brains which help to strengthen neural pathways for all types of learning" (Gainsley, 2011, p. 2).

The math learning you will see in the sand and water area depends on the materials there. Here the children are experimenting with colored water, measuring cups, and a sand/water wheel to see how fast the water goes through it.

What kinds of math experiences will you see in the sand and water area? You might see children

- Compare the size of their sand piles to see which is bigger (KDI 36).

- Count the number of scoops it takes to fill up a container (KDIs 32, 37).

- Hide small animals under bird seed, race cars through tubes, and pour gravel down chutes (KDI 35).

- Make imprints with sifters and notice the design or line up measuring cups by size (KDI 38).

- Build cylindrical sandcastles and describe dug-out roads as *curvy* or *straight* (KDI 34) (Gainsley, 2011).

The following anecdotes include some other things you might see children doing in the sand and water area.

KDI 32. Counting

Mason discovers that there are plastic insects, worms, and butterflies buried in the sand. He picks out some of the plastic insects and places them along the edge of the sand and water table while saying "One, two, three, seven, nine."

KDI 33. Part-whole relationships

Diego brings the tub of small dinosaurs to the sand table that has a three-story graduated box structure. He places several dinosaurs (one handful at a time) on each level of the structure.

KDI 35. Spatial awareness

One day the children scoop so much sand to one corner of the sand table that part of the clear bottom of the table is visible. Avery says, "Look, Henry is under the table waving at us."

KDI 36. Measuring

Olivia holds up two clam shells she finds in the sand table and says, "These are the same, but this one is big and this one is little."

KDI 38. Patterns

Fatima moves the dirt around looking for gems to use to make a path to her castle. She starts the path with a green one, then puts down a purple one, then a blue one and alternates them, green-purple-blue all the way to the castle. "See, Dom, my castle has a princess sidewalk," she says.

Math in Action: Sand and Water Area

The contents of the materials in sand and water area will affect the nature of the play you observe there. For example, play with discrete materials, such as pebbles and gems, might involve counting or part-whole relationships (combining and separating sets of objects). Play with continuous materials, such as sand or water, might include explorations of volume. Both types of materials can lead to comparing quantities. Because of this variation, we offer three shorter scenarios that take place in the sand and water area, each involving a different material and/or toy. In addition, while all three scenarios illustrate how teachers support children in the classroom, there are also many opportunities for families to support children at home. Therefore, in the third scenario, instead of following the scenario with support strategies, we describe how the teacher helps parents see all the math learning their child is engaged in during a particular work time.

Wyatt and Aisha moving water

There is green water on both sides of the sand and water table today. A wooden bar, which runs the length of both sides of the table, holds a variety of funnels and several PVC pipes in its holes. In addition, there are a number of measuring cups, plastic wide-mouthed bottles, bowls in several sizes, and poultry basters.

Wyatt and Aisha each have a measuring cup and several of the plastic bottles. They begin scooping up water with their cups and pouring it into their bottles. Wyatt scoops so fast that he doesn't get much water into his cup and then, in his hurry to pour it into the bottle, he spills even more water. Jill (a teacher) comments, "It's hard to get all the water into the small hole in the bottle. It's spilling down the side." Wyatt explains, "I was trying to go super-fast" (KDI 36).

Wyatt fills the cup again and pours water into the bottle, but still spills a fair amount down the side. He says to Jill, "Most of it is still going down the side" (KDIs 32, 35). "Hmm," says Jill, "I wonder if there is something else you could do so more of the water ends up in the bottle." Wyatt looks around and sees a funnel, which the children used recently during a small-group-time activity. "I know!" Wyatt says. "I can put the funnel in the bottle to make sure that the water goes straight in" (KDI 35). Jill smiles and says, "That's an idea. The top of the funnel is wider than the bottle hole. You can pour water into the funnel and it won't spill down the side."

Meanwhile, Aisha picks up one of the poultry basters and says, "I'm going to count how many squirts it takes to fill up the bottle" (KDI 37). *Jill says to her, "Aisha, I am very curious to discover how many squirts it will take." Aisha fills the baster with water and starts counting* (KDIs 32, 37). *She continues adding water to the bottle. Jill watches and points out to Aisha that sometimes the baster is not all the way filled with water. Jill asks, "Is it okay that sometimes the baster is full and sometimes it only has a little water in it? Do they both count as a squirt?" Aisha responds, "Yes, that doesn't matter," and she continues to fill and count* (KDIs 32, 37). *Jill picks up a baster and begins filling her own bottle and counting.*

Support strategies

- Jill uses spatial concepts as she comments on the size of the bottle opening ("It's hard to get all the water into the small hole in the bottle. It's spilling down the side") as she describes Wyatt's problem with filling the bottle.

- Urging him to think of some alternatives, Jill says, "I wonder if there is something else you could do so more of the water ends up in the bottle."

- Acknowledging Wyatt's idea to use the funnel, Jill compares the size of the bottle opening with the size of the funnel: "The top of the funnel is wider than the bottle hole. You can pour water into the funnel and it won't spill down the side."

- As Aisha fills the bottle with water, Jill introduces the idea of using a standard unit of measurement: "Is it okay that sometimes the baster is full and sometimes it only has a little water in it? Do they both count as a squirt?" When she sees that Aisha is more interested in counting than using a standard measurement, Jill uses the materials in the same way Aisha does.

Dom and Mason cooking in the dirt

Today there is some rich, dark dirt on both sides of the sand and water table. Both Dom and Mason make a plan to cook at the dirt table. They head toward the house area to get the pots that they will use with the dirt. They fill their pots with dirt, stir, and pretend to give each other a taste from their pots. They make faces and say "yuck" as they taste each other's cooking.

As Elena (a teacher) walks by the table, Dom invites her to taste his cooking. Elena pretends to taste the dirt, makes a face, and asks, "What did you make?" Dom replies, "Dirt stew." "It tastes different than anything I have ever tasted," says Elena. "How did you make it? I'd like the recipe." Dom replies, "You put in three cups of dirt and some salt and some eyeballs" (KDIs 31, 36). *"Oh, wait," says Elena. "I want to write this down." She uses her notepad and writes down:*

— 3 cups of dirt
— Salt

Then she asks, "How many eyeballs?" Dom replies, "Oh, about a thousand" (KDI 31). *"Oh, that's a lot. I'll have to write the number 1000 on my paper," Elena says. She writes 1000 as Dom looks on. Dom says, "Or you can just put in six"* (KDI 31). *Elena writes the numeral 7 on her paper and shows it to Dom. "No," says Dom. "You made a 7. A 6 goes like this"* (KDI 31). *He writes the numeral 6 in the air with his finger* (KDI 31), *and Elena writes a 6 on her paper.*

Elena then gets a pot from the house area and picks up a measuring cup from the dirt. "Dom, do I need a large or small measuring cup for the dirt?" she asks. Dom looks at the cup she has in her hand and says, "Bigger" (KDI 36). *Elena asks Mason if he has a bigger measuring cup for her to use. He hands her a bigger cup* (KDI 36). *Elena stirs the dirt in her pot and then says, "I need some eyeballs. I need six. Where are the eyeballs?" Mason picks up a clump of dirt and says, "Here are eyeballs." He puts the dirt into Elena's pot. "Oh thanks, Mason. I need six eyeballs," Elena says. Mason picks up another clump of dirt, puts it in the pot, and looks at Elena. "Thanks, Mason. That's two eyeballs in the pot. I need more," Elena says. Mason continues to put clumps of dirt into the pot and Elena counts each time he adds a clump until they reach six. "This is going to be delicious stew," says Elena.*

Support strategies

- Elena models writing numerals as she writes down Dom's recipe (3 cups of dirt, Salt).

- When Elena asks Dom how many eyeballs she will need for the recipe, Dom replies, "Oh, about a thousand." Elena's response that one thousand is a lot encourages him to think about quantity.

- After Elena writes 1000, Dom tells Elena that she can put in fewer eyeballs: "Or you can just put in six." Elena writes the number 7 on her paper and shows it to Dom. "No," says Dom. "You made a 7. A 6 goes like this" and writes the numeral 6 in the air. By intentionally making this error as she writes down the recipe, Elena is also finding out if Dom recognizes certain numerals.

- Elena introduces measurement concepts when she asks Dom which size of measuring cup she should use: "Dom, do I

Today the sand and water table contains dirt. During work time, the children don construction hats (from the house area) and use toy construction vehicles (from the toy area) and the sand/water wheel to do construction work.

need a large or small measuring cup for the dirt?"

- As Mason drops the clumps of dirt into the pot, Elena models counting.

Mina and dinosaurs

Today there is water on one side of the table and fine play sand on the other side. Mina is playing alone today, wholly absorbed with the dinosaurs. Mina begins pushing the sticks into the sand, counting to herself, "1, 2, 3, 4, 5, 6 — 6 trees. I need more than that for a jungle" (KDI 32). *She goes over to the art area to get more. When she returns, she notices that some of the trees are leaning over and continues talking to herself, "They're going to fall down. The sand isn't strong enough." She gets a cup of water from the other side and mixes it in with the dry sand. She murmurs, "That's better"; sticks in four more sticks (trees); and begins to count all the trees, touching each one as she counts: "1, 2, 3, 4, 5…" She continues until she gets to the last one and says, "10! I have 10 trees!"* (KDI 32). *She places some rocks and stones around and then chooses several of the dinosaurs. She pretends that they are walking over the stones, toward the jungle.*

Jill (a teacher) kneels down next to Mina and comments, "You have lots of sticks poking up." Mina replies, "I have 10 trees" (KDI 32). *Jill pokes some sticks into the sand next to her. Mina picks up a dinosaur and knocks over Jill's sticks with its tail. She says, "This dinosaur knocks over trees with its long tail"* (KDI 36). *Jill repeats Mina's idea by saying, "That dinosaur has a long tail to knock over trees. All my trees are knocked down." Jill counts, "1, 2, 3, 4. Four trees down on the ground, and your 10 trees standing up straight."*

Mina knocks down some of her sticks and looks at Jill. Jill shrugs her shoulders and waits for Mina to say something. Mina counts the sticks, saying a number and touching a stick but counts some of the sticks more than once, "1, 2, 3, 4, 5, 6, 7, 8, 9, 10, 11, 12…13 trees knocked down" (KDIs 32, 35). *Jill says, "I'm going to line up the trees so it's easier for me to count." She lines up the sticks that have fallen and counts out loud, "1, 2, 3, 4, 5, 6, 7, 8, 9, 10, 11…11 trees down."*

Yesterday, while they were making their plan for today, Jill and Elena talked about Mina, since they were also going to be having a family conference with her parents later this afternoon. Jill made notes as she played with Mina, preparing comments for Mina's parents. When Mina's parents arrived, Jill described what she observed Mina doing while she was playing with the dinosaurs at the sand table during work time, stressing the math learning that occurred. Here are some of her comments:

- Before I joined Mina, I observed her first and I heard her counting as she pushed sticks into the sand. When she completed counting, she repeated the last number counted. She said, "1, 2, 3, 4, 5, 6 — 6 trees." This tells me that Mina understands cardinality; that is, that the last number counted is the quantity.

- Mina was having a hard time getting the trees to stand up in the loose sand, so she went and got some water to mix in. When the mixture was the way she wanted it, she added four more trees to the six still standing. Once again, Mina demonstrated her understanding of cardinality as she counted the trees, touching each one as she counted, "1, 2, 3, 4, 5…" until she got to the end and said, "10. I have 10 trees!"

- I also put some trees in the sand, and Mina brought her dinosaur over and used his tail to knock over some of my trees. She explained it by saying that the dinosaur knocks things over with its "long tail." When Mina described the dinosaur's tail as "long," she was using a measurement term.
- Finally, I noticed several times throughout our play that Mina has an understanding of a number of spatial awareness terms. For example, Mina made the following comments:

   "They're going to fall down."
   "This dinosaur knocks over trees...."
   "...13 trees knocked down."

Over There! Catch the Math in the Toy Area

Math learning opportunities are in virtually every storage container and on every shelf in the toy area (see pp. 38–41 for additional information on toy area materials). Children build, knock down, balance, add on to, and take away from structures in the toy area, using construction materials like Magna-Tiles (which come in a challenging variety of shapes and sizes), Tinkertoys, and Lincoln Logs. Small, stackable manipulatives such as pegs (and pegboards), Unifix cubes, linking cubes, and small blocks often produce conversations that will include counting, measurement (*smaller, tallest, biggest*, etc.), and pattern recognition terms (e.g., "It's a pattern. You need a blue one next").

From a mathematics perspective, you might say that the toy area is the "total package," providing ready opportunities for children to experience any one of the nine math KDIs. As you read the anecdotes and scenarios in this section, imagine you are in your classroom and think about how you might scaffold the children's math learning that is taking place.

KDI 32. Counting

Dom puts pegs in a pegboard and says random numbers as he does so. When he finishes, he says, "I got many of them."

KDI 32. Counting
KDI 33. Part-whole relationships

Henry counted out 10 Lego people from the bin and divided them into two sets. He points to each pile as he says, "There are four guys here and six over here."

KDI 34. Shapes

Sabrina is building a house with Magna-Tiles. She puts four small squares together and says, "If you put squares together, it makes another square."

KDI 35. Spatial awareness

Wyatt looks at a puzzle piece he is holding and says, "It's upside down."

KDI 36. Measuring

Aisha is building a tower with the Magna-Tiles and when Elena comes over, Aisha points at the tower and says, "You are taller than that."

KDI 36. Measuring

Jamee is watching Mina and Razi stack the nesting blocks. When they finish, she points to the block at the bottom and then, proceeding to the top block, says, "Huge, big, medium, small, almost tiny, tiny."

Math in Action: Amare Plays With Inch Blocks and Small Reptiles

In this scenario, Elena (the teacher) joins Amare during work time and observes him before commenting on what he is doing. Let's see what math KDIs Elena observes and look at the support strategies she uses.

Amare leans against the shelves in the toy area with the tub of inch blocks and a basket of small lizards and snakes beside him. He takes out several handfuls of blocks and dumps them on the floor in front of him. He moves them around so one group has assorted color blocks and the other group has mostly white blocks (KDI 33). While he is looking through the basket of lizards and snakes, Elena (the teacher) joins Amare on the floor, silently watching to see what he is going to do next. He finds two lizards and places each one on the top of a white block (KDI 32). He continues looking, and each time he finds a lizard, he puts it on a white block until each white block has a lizard on it (KDI 32). Elena comments, "Amare, you put one lizard on each white block." She points to each block and counts, "One, two, three, four, five lizards on five blocks." Amare looks up at her and smiles. He says, "Now I'm going to put snakes on the colored blocks" (KDI 32). Elena responds, "I think I'm going to try putting some lizards on some of the blocks too." She reaches for the lizards and blocks and says out loud, "One, two, three lizards and one, two, three blocks."

In the toy area, you might find children combining and separating collections of materials, such as shells and small figures and animals.

Support strategies

- Without interrupting him, Elena joins Amare and watches him while he puts the lizards on the blocks. When Amare finishes, Elena uses math language to describe what she saw him doing: "Amare, you put one lizard on each white block." She then adds a gentle extension: "One, two, three, four, five lizards on five blocks" to support Amare's developing understanding of one-to-one correspondence.

- When Amare says he intends to do something similar with the snakes and the colored blocks, Elena decides to stay and use the materials in the same way: "I think I'm going to try putting some lizards on some of the blocks too." She reaches for the lizards and blocks and says out loud, "One, two, three lizards and one, two, three blocks," further reinforcing the idea of one-to-one correspondence.

Math in Action: Kyra, Magna-Tiles, and Peach Pie

During this work time, Jill (the teacher) supports Kyra's understanding about shapes so that Kyra can achieve her goal: to make a peach pie.

"Jill, Jill, can you help me?" Kyra asks as she holds up two different Magna-Tile shapes. Kyra continues, "These all have three sides, but I need the ones that look like pie!" (KDI 34). Looking slightly confused, Jill (a teacher) says, "What do you mean?" Kyra explains, "They need to be long and skinny" (KDI 35). Jill holds up a triangle and says, "Oh, the sides of this triangle are all the same. It's not the kind you need?" Kyra shakes her head. Jill holds up another one and asks Kyra, "What about this one?" Kyra takes it (and several others like it) and tries to fit them together so they make a circle, but it doesn't work. She frowns and says, "No, they have to make a circle like a pie" (KDI 34). Jill holds up another triangle and says, "How about this one; it has two long sides and the third side is short." Kyra tries to put them together by flipping them in different ways until she notices that they fit together like pie pieces (KDI 34). "These are the right ones," she exclaims to Jill, "I think they'll make a whole circle" (KDI 34). She then collects additional pieces to complete her pie. Kyra fits in the last triangle and uses the small yellow counting bears as peaches, counting as she places three or four on each piece (KDIs 31, 32, 34).

This child manipulates triangle Magna-Tiles to create a circular pie.

Support strategies

- Jill supports Kyra's understanding of triangles by acknowledging Kyra's statement that triangles have three sides and showing an example: "Oh, the sides of this triangle are all the same. It's not the kind you need?"

- Jill also uses the math term *third* in a new context for Kyra. Normally (e.g., on the message board), Kyra is accustomed to hearing *third* follow *first* and *second*; however, this time, instead of saying the "first and second long sides" which is rather cumbersome, Jill simply says "two long sides and the third side...."

- Having found the kind of triangle Kyra needs, Jill stays to watch Kyra flip and turn the pieces until they fit. After helping Kyra find the correct triangular shape to make her pie, Jill supports Kyra by being available for help but does not interrupt Kyra by constant offers of help or distract her with questions.

Math in Action: Diego and Avery Weigh Shells and Stones

"Let's see how much that big shell weighs using the stones," Diego says to Avery (KDI 37). *"You mean the really big one that Elena brought in that's bigger than all the others"* (KDI 36)? *asks Avery. "Yeah, it's big, but I don't think it's as heavy as a big rock"* (KDI 37), *says Diego. They get out the balance scale, the basket of shells, and the basket of stones. "Good, the big shell is in here," says Diego, as he places it in a bucket on one side of the scale* (KDIs 35, 36). *Then Avery says, "Let's start putting rocks in the other side, and we can count how many when the scale says it's the same"* (KDI 36). *As they are adding stones to the bucket on the other side of the balance, Elena (a teacher) joins them on the floor. "Elena, see we're weighing your giant shell," Diego exclaims. "I noticed that; that's why I came over. You are putting the stones on the other side of the balance. It's a kind of scale, a way to weigh things," she explains. "Yup," says Avery. "Then when the sides are the same* (KDI 36), *we'll pour out the stones and count them." Elena observes, "You've already put some stones in there; I wonder how many more you will need?" The boys look at the stones in the basket, and then Avery says, "Oh, maybe five or seven"* (KDI 32).

They continue to add one stone at a time, checking the scale each time to see if the buckets are nearly even. "I think if I put this one in, it will be just right," says Diego as he places it gently with the rest of the stones. "Oh no," says Avery, "take it off; it's too heavy" (KDIs 37, 36). *"How could you tell it was too heavy, Avery?" asks Elena. Avery explains, "When the stone was on there, the stone bucket went down lower than the bucket with the big shell. That means it's heavier"* (KDIs 35, 37). *"I wonder if there is a way to get it to balance — to make it so the buckets weigh the same," muses Elena. Diego looks through the basket of stones and picks out several. He holds them out and says, "We could try one of these; they are smaller than the last one"* (KDIs 31, 36).

Support strategies

- Elena joins the boys on the floor without interrupting their activity. By simply observing, she demonstrates respect for what they are doing and allows them to retain control over the play situation.

- The boys tell Elena what they are doing and she restates what they said, introducing new vocabulary by using the word *balance* instead of *scale* and explaining that a *balance* is "a way to weigh things."

- Elena offers Avery and Diego a gentle extension, encouraging them to verbalize what they are thinking about and to estimate how many more stones they will need: "You've already put some stones in there; I wonder how many more you will need?"

- As soon as Diego puts on the last stone, Avery tells him to take it off because it is too heavy. Elena asks Avery to explain what made him think it was too heavy ("How could you tell it was too heavy, Avery?"), knowing that by doing so she is encouraging him to think through his reasoning in more detail.

- Following Avery's explanation, Elena provides a gentle extension by saying "I wonder if there is a way to get it to balance — to make it so the buckets weigh the same." She states what she means by *balance* to ensure that the boys will understand.

Over There! Catch the Math in the Book Area

In the book area, children look at and read books to themselves, to each other, to Valentine (the class pet), to an adult, and, of course, listen as an adult reads a book to them. It is a cozy spot with soft furniture and pillows where children can relax as they look at the books or magazines available to them. Sometimes children will go to another interest area to find props to supplement a story if they are re-enacting or retelling a story in their own words. Other children may choose to write and illustrate their own stories. Even though the book area is for reading, there are ample opportunities for math learning experiences as demonstrated in the following examples.

KDI 31. Number words and symbols

Jill (a teacher) and Olivia are reading The Three Bears. *As Goldilocks looks into the room upstairs, Olivia says, "I know, three beds. Like three chairs, three bowls, and three bears."*

KDI 34. Shapes
KDI 36. Measuring

Theo and Peter sit on the couch in the book area and look at the wordless picture book Changes, Changes, *which is about a wooden couple who live in a building-block house that suddenly catches fire. Theo explains to Peter, "The arch block is on fire, so they decide to build a fire truck to put out the fire. To get started, she carries some of the small cylinder blocks and he carries the long, rectangle block."*

KDI 35. Spatial awareness

David (a teacher) reads the book Quick As a Cricket *to Molly and Tania. When they get to the page that says, "I'm as shy as a shrimp," Molly points to the picture and says, "The little boy is hiding inside the shell."*

KDI 35. Spatial awareness
KDI 36. Measuring

Lucy, Lexie, and Na'treal look at one of their favorite books, Brontorina, *about Brontorina, a dinosaur who wants to be a ballerina. They show the book to Mrs. Mac (a teacher). At a particular page, one of the girls turns to her and says, "This is sad. The teacher says Brontorina is too big to be a ballerina. Yeah, and they can't find a dancer to lift her up high."*

Math in Action: Randy Reads The Very Hungry Caterpillar

Work time can also include time curling up with a good book. Let's see how this teacher supports a child's mathematical understanding while reading *The Very Hungry Caterpillar*.

Randy brings The Very Hungry Caterpillar *over to Elizabeth (a teacher) and asks if she would read it to him. They snuggle up on some of the pillows, and Elizabeth begins to read. When they get to the page where the caterpillar begins to look for food, Randy says that he wants a turn to read and turn the pages. On one page, he points to each piece of fruit as he counts. "He ate one apple, but he was still hungry, then he ate two pears — still hungry! Then he ate three plums but still wanted more, then he had four strawberries and he was still hungry. Finally, the next day he had five oranges, and guess what" (KDI 32)? Randy asks as he looks at Elizabeth. "He ate lots of fruit?" asks Elizabeth. "Nope," says Randy, "He ate one apple, two pears, three plums, four strawberries, five oranges, and he was still hungry (KDI 32). Now you read."*

Elizabeth continues, "Well, since he was still hungry, the next day he ate a lot more," and she reads all the different foods the caterpillar ate on Saturday. She turns to Randy and says, "That seems like so much food! I wonder if we could count them to find out how many pieces he ate." "I don't know," says Randy dubiously. "There are a lot. We could try counting them." "Would you like to count them together?" Elizabeth asks. "Okay," Randy replies, smiling. And they began counting together, "One, two, three..." (KDI 32).

Support strategies

- When Randy asks Elizabeth a question about what happens after the caterpillar eats the next day ("Finally the next day he had five oranges, and guess what?"), Elizabeth uses math language ("lots") to respond, introducing a simple concept of quantity.

- Since Randy was enthusiastic about counting the fruit ("He ate one apple, two pears, three plums, four strawberries, five oranges, and he was still hungry"), Elizabeth decided to offer a gentle extension after reading the long list of foods the caterpillar ate on Saturday: "I wonder if we could count them to find out how many pieces he ate."

- When Randy appears slightly reluctant to count how many pieces of fruit the caterpillar ate, Elizabeth supports him at his level, suggesting that they count the pieces together. Randy readily agrees with this suggestion.

Math in Action: Across the Days and Throughout the Classroom

Talk to your colleagues with classrooms similar to yours — ones in which children have an extended period of time when they are free to pursue their own interests and choose their materials. Are you familiar with those occasions when a group of children develop an intense interest in a play idea and may choose to pursue it for several days or even weeks? Do you also find that these ideas tend to flow beyond the domain of one interest area into multiple areas? As you support children during work time, remember that mathematics learning is not neatly confined to one day, one time of day, or one area. Look for opportunities to encourage children's interest in mathematics across times and places. It will open your eyes and theirs to the many possibilities for exploring all the areas of mathematics during this part of the daily routine.

There are many opportunities for math learning through reading books — including counting, as illustrated in The Very Hungry Caterpillar *scenario on the facing page.*

8

Small-Group Time

There are two aspects to providing occasions for wonderful ideas. One is being willing to accept children's ideas. The other is providing a setting that suggests wonderful ideas to children — different ideas to different children.

— Duckworth (2006, p. 7)

Small-group time is the part of the High-Scope daily routine set aside for children to explore, manipulate, experiment, and solve problems with materials that are part of an activity planned by the teacher. While the adult introduces the activity and materials, children are free to choose how they will use the materials and describe the process and their ideas in their own words.

What Children and Adults Do During Small-Group Time

While Eleanor Duckworth's comment above applies to each part of the program day, it is particularly relevant to small-group time. When adults provide interesting and challenging materials interspersed with supportive comments and gentle extensions, small-group time becomes a laboratory for wonderful ideas. Thoughtful and intentional planning on the part of adults are important contributors to a successful small-group time. As we pointed out earlier in this book, this is the part of the day during which children can have "experiences in which mathematics is the primary goal" (NRC, 2009, p. 2). However, even in small-group times in which math is not the primary focus of the activity (e.g., a literacy, science, or art activity), adults need to be alert to the math learning that can happen when children explore other content areas.

In chapter 4, we explored strategies and ideas for supporting children and their learning throughout the day. Because teachers sometimes use small-group time to introduce a new concept, it is important to briefly review the teacher's role during this part of the daily routine. Once the children have begun to work with the materials, teachers pay attention to their actions and ideas, scaffold further learning, and encourage them to interact with and learn from one another by

- Closely attending to each child
- Getting down on their level
- Watching and listening to them
- Imitating and building on their actions

- Conversing with children and following their leads, asking questions sparingly
- Encouraging children to solve problems
- Referring children to one another for ideas and assistance (HighScope Early Childhood Staff, 2009)

Over There! Catch the Math During Small-Group Time

One of the reasons it is important to keep your math senses sharp and well tuned is because math learning may occur when you least expect it. For example, be alert for math occurring during small-group times even when you had something other than math in mind, as illustrated in the anecdotes here.

> KDI 31. Number words and symbols
> KDI 32. Counting
> KDI 33. Part-whole relationships

Mr. Vince is doing a small-group time in which he plans to focus on KDI 21. Comprehension. He begins by reading the story of The Three Little Pigs. *He has small animals and small construction blocks for each child. As the children start playing with the materials, they make comments to him and to one another:*

- *"I got three elephants," Jake says to Mr. Vince, as he points to each elephant.*
- *"I need more to finish the third house," Maya says to Shara.*
- *"Hey, Mr. Vince," Finn says, pointing to a pile of animals in front of him. "I had two cows, two pigs, and one horse. Now I have five animals!"*

During small-group time, the teacher introduces the activity and provides the materials. The children then choose how they will use those materials.

KDI 31. Number words and symbols

Taro says "Many" after putting pegs in a pegboard and saying random numbers in Spanish.

KDI 34. Shapes
KDI 35. Spatial awareness

Since several of the children in her group have new babies in their families and there has been a lot of baby play lately, Ms. Angela has baby dolls, baby clothes, blankets, and diapers for each child during this small-group time. Here are some of the math experiences her children encounter:

- *Dashawn holds up a onesie and says, "This is baby's underwear. It goes under."*
- *Jamilla struggles with a diaper, and Ms. Angela offers to help. Jamilla gives Ms. Angela a piece of tape and says, "I'm holding this side down while you tape it."*
- *Miriam says to Juliet, "I think I'm going to put the shirt with the pink triangles on my baby."*

KDI 31. Number words and symbols
KDI 36. Measuring
KDI 37. Unit

Rosario notices that several of the children in her group have been building in the block and woodworking areas. She is interested in learning what they will make with the new materials. Today, to capitalize on their interest in building, she provides each child with a variety of materials, including plastic containers, Styrofoam pieces, foam shapes, Popsicle sticks, and cardboard tubes. The children use the materials in different ways:

- *Audrey stacks three plastic containers on top of each other and wants to add another but she can't reach the top. She says to Rosario, "Can you help me? Then it'll be the tallest! See it's bigger than me now!"*
- *Charlie and Jesse work together on a "rocket ship." "Is it big enough yet?" asks Jesse. "We should measure it," says Charlie. "Hold the tape measure and we'll see." Jesse holds the tape measure at the bottom of the structure, and Charlie stretches it to the end. "It says 18," says Charlie. "If we add one more tube, it'll be long enough."*
- *Coretta picks out large- and medium-sized containers and decorates each one. First she pats the top of the large one and then she tries the medium one. She turns to Shara and says, "I made drums! The big one is louder than the smaller one. You want to hear?"*

KDI 38. Patterns

It is early in the school year, and Cheryl plans a small-group time working on fine-motor skills. Each child receives his or her own basket with some shoelaces, pipe cleaners, and wooden beads. The children in Cheryl's small group use the materials in different ways, as illustrated here:

- *Leah puts three red wooden beads on her shoelace, adds a large blue bead, and then repeats the pattern. She holds it up to Cheryl and says, "See, I made a necklace, it has three red ones, then one big blue one, then three more red ones, one big blue one, and then three more red ones. Cheryl comments, "I see a pattern in your necklace — it goes three red beads, one large blue one; three red beads, one large blue; three red beads, one large blue." She scaffolds Leah's math learning by naming what she did as a pattern and then describing what made it a pattern.*

During this small-group time, the teacher provides the children with different-colored plastic insects. She scaffolds this child's learning by asking the child how many spiders are the length of an inchworm.

- *Maya picks out a green pipe cleaner from her basket. She then picks out all the yellow and green beads from the basket. Cheryl sits next to her and gets some materials for herself and begins to imitate Maya. Maya begins to put the beads on the pipe cleaner, alternating a yellow bead and a green bead the length of her pipe cleaner.*

Among preschool-aged children, you are more likely to see data analysis taking place in a group, rather than individually in a small-group time. However, the following are some examples that might occur during a small-group time.

KDI 39. Data analysis

Taro takes all the frogs out of his basket and lines them up one behind another. Next he takes out the long lizards and lines them up the same way, next to the frogs. He then takes the beetles and lines them up next to the lizards. He points to the line of frogs, which is the longest, and says, "Las ranas más [more frogs]."

KDI 39. Data analysis

Finn looks around the table and then over at the other table. He concludes, "There are more kids at our table than at Ms. Angela's."

In the remaining part of this chapter, we will look at two small-group times that illustrate the kind of math experiences that can take place during this part of the daily routine. Following each small-group-time example, we will review the strategies used by the teachers to support children's math learning.

Math in Action: Golf Tees and Styrofoam

Coteachers Ms. Angela and Mr. Vince plan on bringing the woodworking table into the classroom next week. To introduce the children to the tools they will find there, Ms. Angela has large chunks of Styrofoam and lots of golf tees in baskets for each child. She also has a hammer for each child as a backup material.

Miriam

Ms. Angela sits down with a basket of materials and tells the children about the woodworking table as she introduces the materials. Once the children each have a basket, they get very busy, some trading golf tees to get certain colors while others quietly focus on getting the tee to go into the Styrofoam.

Ms. Angela gets up and sits down quietly beside Miriam, who has just finished counting a pile of tees. "Ms. Angela, I only had three pink tees, but DeClan had three pink ones too, so I gave him three green ones and he gave me the pink ones" (KDI 32). "It sounds like you guys were solving a problem," says Ms. Angela. "Now I have more pink ones to use and pink is my favorite" (KDI 32), *says Miriam.* "Now that you have three more from DeClan, is that enough for what you want to do?" *asks Ms. Angela.* "I'll count them," *says Miriam. She counts them and says,* "Yes, I have six pink tees now" (KDI 32).

Support strategies

- By silently joining Miriam, Ms. Angela gives Miriam the opportunity to initiate the explanation about trading with DeClan for the pink tees.

- Ms. Angela acknowledges Miriam's solution to her predicament of too few pink tees in a way that worked for DeClan too: "It sounds like you guys were solving a problem."

- Ms. Angela is curious if Miriam knows how many pink tees she has now: "Now that you have three more from DeClan, is that enough for what you want to do?" This statement/question is phrased in a way that gently suggests that Miriam might want to count the tees to see if she has the number she needs for what she wants to do.

Charlie

Charlie gestures to the tees remaining in the basket and says to Ms. Angela, "I need more tees. I put almost all mine in [the Styrofoam]" (KDI 33). "Charlie, there are more tees in that yellow basket," *Ms. Angela replies. Charlie gets another handful and picks up one to stick in the Styrofoam. Suddenly he says,* "Hey, Ms. Angela, look at the top of this yellow tee — there's a 2 on it" (KDI 31). *Ms. Angela looks closely and says,* "Charlie, I can see the 2. I wonder why there is a 2 on it." "I don't know," *he responds.* "Maybe different colors have different numbers? I'm going to look through the rest of these and see if I can find any more numbers," *he says.* "Okay," *says Ms. Angela.* "I'll be back to find out what you discover. I am curious to see what you find out," *she adds.*

Support strategies

- Ms. Angela confirms that Charlie found a 2 on the golf tee and wonders out loud why it's there. By reflecting out loud about the presence of the number on the golf tee, Ms. Angela is inviting Charlie to think about whether or not there might be numbers on other different-colored tees.

- When Charlie, diverted from poking the tees in the Styrofoam, decides to look for additional numbered tees, Ms. Angela supports his choice ("I'll be back to find out what you discover") and also lets him know that she is genuinely interested ("I am curious to see what you find out").

Derek

Ms. Angela joins Derek, who is holding a tee and poking it in and out of the Styrofoam while saying "One" with each poke (KDI 32). Ms. Angela gets her own piece of Styrofoam and a tee and imitates Derek's movements, saying "One" each time she sticks her tee in. Derek looks at her and smiles broadly; she returns his smile and continues with the same action. Derek also continues with what he is doing. After a while, Ms. Angela picks up a second tee. She sticks one tee into the Styrofoam and says, "One." Then she sticks in the second tee and says, "Two." She continues to do that until Derek starts watching her. She slows down, holds up a tee, says "One" as she pokes it in the foam, then picks up the second tee and says "Two" as she pokes it in. She repeats this several times, keeping her movements slow.

Support strategies

- Ms. Angela has her own piece of Styrofoam and some tees so that she can imitate what Derek is doing and build on his action to support his math learning.
- After they have both have been sticking one tee in the Styrofoam (each saying "One" with the action) for a while, Ms. Angela picks up a second tee and begins alternating her actions and number words: sticking in the first tee and saying "One" and then the second tee as she says "Two."

- Ms. Angela adds the word *two* to her comment very casually; she does not want to disrupt Derek's play because she knows that if she has a previously determined agenda, she is not as likely to have success in engaging him.

- When Derek begins to watch Ms. Angela, she slows down her movements so he will be able to follow her.

- In addition to introducing number words, Ms. Angela is helping Derek see the one-to-one correspondence between each tee and the number word she attaches to it.

- Ms. Angela repeats the words *one* and *two* several times, always keeping her movements slow and letting Derek watch her mouth as it forms the words.

Math in Action: Shape Blocks

The shape blocks (i.e., blocks in different colors, shapes, and sizes) are a new material in the classroom. After he introduces them to his group, Mr. Vince passes out baskets of the shape blocks to each child and then takes some blocks from his own basket and lays them out on the table. He turns some squares on their sides and puts a diamond on top of one and a parallelogram on top of another.

Jo-Jo

Mr. Vince then turns to Jo-Jo, who is sitting next to him and piling the blocks on top of each other. "Jo-Jo, you discovered that you can stack these blocks," says Mr. Vince. "Yeah," says Jo-Jo, "and I found out that it won't fall down if I put the big, yellow ones on the bottom" (KDI 35). Mr. Vince starts

During this small-group time, the children use the shape blocks on the carpet. After making a very large design, this child measures himself against his creation.

to imitate Jo-Jo's tower. "After you put the yellow hexagons down, you stacked the squares next and then the triangles on top? I want to do it the same way with the same number of yellow hexagons first. How did you decide it was best to start with the yellow hexagons?" Mr. Vince asks. "Well," Jo-Jo explains, "they are the biggest; when I started with the squares or triangles, the tower kept falling. I think they're too little" (KDIs 34, 36). *"I used all my yellow hex…"* (KDI 34), *Jo-Jo hesitates at the unfamiliar word. "Hexagon," Mr. Vince adds. "They're the shapes with six sides. It looks like you have quite a few." "I counted," says Jo-Jo. "I have eight* (KDI 32). *I'll show you." Jo-Jo starts at the bottom, carefully pointing to each hexagon as he counts to eight. "Eight* (KDI 32)*," he finishes. "You have eight yellow hexagons," smiles Mr. Vince, "so I'll need eight too." He counts four and says, "I only have four. I need more." Jo-Jo gathers more and counts, "Five, six, seven, eight"* (KDI 32) *as he adds each block to Mr. Vince's pile.*

Support strategies

- When Mr. Vince turns to Jo-Jo, he makes a simple observation about what Jo-Jo had done with the shape blocks: "Jo-Jo, you discovered that you can stack these blocks." Opening an interaction with such a comment lets Jo-Jo know that Mr. Vince is curious about what he is doing and would like to know more.

- Mr. Vince knows that most of the children are unfamiliar with the names of the shapes they don't frequently see (e.g., hexagon, trapezoid, parallelogram, etc.), so he uses the term *yellow hexagons* when he asks Jo-Jo how he decided to use them.

- Jo-Jo tries to use the term *hexagon* as part of his explanation but gets stuck on the word, so Mr. Vince supplies him with the rest of the word and briefly defines it as a shape with six sides.

- At another time, Mr. Vince might suggest counting the number of sides in the hexagon together with Jo-Jo, but he rightly judges that, in this case, it would interrupt the flow of Jo-Jo's play.

- Although Mr. Vince recognizes that it is not the moment to drill Jo-Jo on the meaning and pronunciation of hexagon, he makes a mental note to add the big hexagonal nuts and bolts to the toy area, which will give the teachers another opportunity to use the term with the children.

- Mr. Vince then observes that Jo-Jo has "quite a few" of the yellow hexagons. Included in that statement is a gentle extension with a subtle suggestion that Jo-Jo count his hexagons.

- After Jo-Jo counts how many hexagons he used, ending with "Eight," Mr. Vince offers a summary statement ("You have eight yellow hexagons") to emphasize that the last number counted tells "how many."

- Mr. Vince counts four hexagons in his pile and comments, "I need more," creating an opportunity for Jo-Jo to find additional hexagons and count them, which he does.

Jamilla

Mr. Vince sits down by Jamilla. "Mr. Vince, I made a shape wall; it's almost all the same, except when I ran out of the yellow hexes and had to use a different shape — the diamonds worked okay then" (KDI 34). *Mr. Vince studies Jamilla's wall and says, "You not only used the yellow hexes — hexagons — and the diamonds, but I see other shapes too. It looks like you had to turn some of the shapes upside down to make them fit together. Some of the triangles are pointing up and some are pointing down." Jamilla explains: "I put the green triangles on the top because they look like hats, and the wall wasn't high enough when the hexes were on the ground"* (KDIs 34, 36).

Support strategies

- Mr. Vince restates Jamilla's description of her shape wall, incorporating her word *hexes* and adding *hexagon* and also pointing out that he sees other shapes.

- Although Jamilla ignores it, Mr. Vince offers a gentle extension when he points out that she made the triangles fit together by turning some of them upside down.

Audrey

This scenario is an example of a recommendation we made earlier in chapter 4 (p. 54): "Practice getting comfortable with silence." Interactions with children do not always require a lot of talk, particularly if they are absorbed in what they are doing. Let's take a look at how Mr. Vince supports Audrey by verbalizing (out loud) his thoughts.

Mr. Vince kneels down by Audrey, who is putting a green triangle on each side of a yellow hexagon. He gets his blocks and begins describing (and doing) what he has seen Audrey do. "You started with one of the big yellow shapes," Mr. Vince says, as he puts one down. He continues talking out loud to himself, "The yellow ones are called hexagons. I also see some green pieces that you have put around the hexagon." Audrey looks up at him and says very quietly, "Mr. Vince, that's a flower and those are triangles" (KDI 34). *Mr. Vince gives her a sheepish look and asks, "Audrey, are the triangles the flower petals?" She nods. "Then can I guess what these are?" he asks, pointing to some green shapes sticking out of the bottom. She nods, beginning to smile. He says hesitantly, "Is it the stem…?" and Audrey adds, "It's made from squares!"* (KDI 34).

Support strategies

- Mr. Vince chooses to use the shape blocks in the same way that Audrey does, without adding the additional pressure of conversation. Instead he "acts" as if he is talking to himself, describing out loud what he is seeing her do and then doing it himself. When it becomes apparent to Audrey that Mr. Vince does not understand what the green pieces represent, Audrey speaks up, explaining, "Mr. Vince, that's a flower and those are triangles."

- As Audrey becomes more at ease with Mr. Vince, she notices that he is unsure of how to describe the "stem" of her "flower." He hesitates a bit ("Is it the stem…?"), giving her the opportunity to complete the description by naming the shapes: "It's made from squares!"

During this small-group time, the teacher supports this child by silently observing how he uses the Cuisenaire rods. Later, she might continue her support by using the materials in the same way he does to encourage him to talk about what he is doing.

9

Large-Group Time

Since ancient times peoples of all cultures have gathered around fires, on hilltops, in clearings, or along the shore to sing, dance, tell stories, and exchange information.
— Hohmann, Weikart, & Epstein (2008, p. 267)

Large-group time is the part of the daily routine when children and adults come together as a whole group to benefit from conversation, companionship, a sense of community, and for the fun of doing things together (Boisvert & Gainsley, 2006). To ensure enjoyable, shared group experiences, adults plan and lead activities that include all five ingredients of active learning (see p. 27 for a more detailed description of active learning).

What Children and Adults Do During Large-Group Time

Moving quickly from one activity to the next, incorporating opportunities for children to make choices and take turns being the leader, and, finally, learning to share control with children contributes to successful large-group times. Of course, this is not possible without intentional planning and thorough preparation on the part of adults.

In HighScope classrooms, large-group times are typically planned around five different types of activities:

- Easy-to-join activities
- Singing, fingerplays, and chants and poems
- Storytelling and reenactment of stories and nursery rhymes
- Movement activities with (or without) music or objects
- Cooperative games

With such a variety of experiences and activities to draw on, many large-group times include a focus on a variety of the curriculum content areas including, of course, mathematics. As we have demonstrated in the previous chapters, however, if you provide the right ingredients (i.e., a mathematically rich learning environment, scaffolding from adults, a regular math focus during small-group times), math experiences will occur throughout the daily routine. Like small-group time, large-group time does not have to have a specific math focus to result in experiences that are rich in math content as you will see in the following examples of large-group times.

Over There! Catch the Math During Large-Group Time

KDI 31. Number words and symbols

When Shelby (a teacher) asks who wants to be the spider in the reenactment of "Miss Muffet," Ben says, "Me first!" followed by Eva, who adds, "Me second!"

KDI 31. Numbers words and symbols
KDI 32. Counting

"One little monkey jumping on the bed. One fell off and bumped her head..." chant the children as Bella pretends to jump on a bed and then fall off. "How many monkeys are left on the bed?" asks Carol (a teacher). "None," says Jamal. "No more monkeys," nods Margie. Carol then asks the group, "What other animals could jump on the bed?" "Dinosaurs," suggests Julia, "lots of dinosaurs, a thousand!"

KDI 33. Part-whole relationships

At large-group time today, the children are going to wave scarves as they move to music. Shelby (a teacher) says, "I have blue, orange, and green scarves." Once all the scarves are distributed, Kamala looks around and says, "There are four kids with green, six with blue, and five kids and two teachers with orange; that's everyone."

KDI 34. Shapes
KDI 35. Spatial awareness

As the music stops, the children find the closest shape (different shapes are scattered around the floor) and stand on it. Carol (the teacher) says, "If you are standing on a circle, wiggle your fingers." Xavier and Blake, who are on circles, wiggle their fingers and start laughing. "I'm on a triangle," says Natreal. Carol asks, "Natreal, how should kids on triangles move?" "Shake their arms over their heads," she responds.

KDI 35. Spatial awareness

Today for large-group time, the class is "popping corn" to "Popcorn Music" by using the parachute to bounce (pop) foam balls. As the music gets faster, Eva says, "This popcorn is crazy! It goes all over the place — under the parachute and even behind the shelves!"

KDI 35. Spatial awareness

"Eva, would you please scoot over closer to Jamal, so Bella has a place to sit?" asks Shelby (a teacher), and Eva moves closer to Jamal.

KDI 36. Measuring
KDI 37. Unit

Marching around the circle, Ben holds his ribbon stick up in the air and says, "Mine is highest." "No, mine is higher, see?" says Austin, holding his stick next to Ben's stick.

KDI 36. Measuring

Some of the children are lying on their backs moving their legs in the air (the teacher has asked the children to "dance in the air" for this easy-to-join activity while the rest of the group finishes their snack). Carol (a teacher) asks them how they would like their legs to dance. "Fast," says Julia. "Like this?" Carol asks, as her legs move around slowly. "No! Faster! Like this," Julia says as she makes her legs go around much faster than Carol's legs.

Math experiences often occur naturally in large-group times. During this large-group time, the children move to music and then stop on a circle once the music stops.

KDI 36. Measuring

As Shelby (a teacher) passes out scarves for the children to use during large-group time, Jackson says, "I want the longest one."

KDI 38. Patterns

Margie looks around the circle and says, "Look! We're standing boy-girl-boy-girl-boy, at least until we get to Lee and his dad."

KDI 38. Patterns

Carol (a teacher) starts patting her shoulders and then her head and repeats the movement pattern while the children imitate her. "Who has an idea of two different places to pat?" she asks. "Tummy and nose," suggests Luis. "This way," he says, as he begins patting his tummy and then his nose, repeating the pattern until it is someone else's turn to choose.

KDI 38. Patterns

The children march around the room, each holding a bright-colored feather duster. It is Maykayla's turn to be the leader, and she lifts her duster up and down while saying "Up-down-up-down-up-down."

KDI 39. Data analysis

"More kids are jumping than hopping," says Xavier as he observes how the children have chosen to begin this easy-to-join activity.

KDI 39. Data analysis

At the end of large-group time, Natreal comments, "Jamar, Danielle, Joanie, and Albert got to be Miss Muffet. All the other kids got to be the spider. Lots more kids like to be the spider!"

At this large-group time, the children use a large parachute to "pop popcorn" (foam balls) slower and then faster to "Popcorn Music."

Math in Action: The Three Billy Goats Gruff

Shelby and Carol (the teachers) hope to observe the children experience both music and math KDIs during large-group time today. They decide to tell the story of *The Three Billy Goats Gruff* and give the children sets of rhythm sticks so they can help to tell the story. The teachers have chosen *The Three Billy Goats Gruff* because the narrative offers them opportunities to use math-related language with the children.

> *Shelby begins, "Today we are going to tell you a story about some hungry goats that decide to go up the hill so they can fill up on their favorite grass, which only grows in a special place on the hill." A number of the children shout out, "The Three Billy Goats Gruff!" Shelby nods and says, "That's right! It's the story of* The Three Billy Goats Gruff, *and we noticed that some of you have been looking at the book during greeting time." As Carol passes out the pairs of rhythm sticks to each child, Shelby explains that the children can use the rhythm sticks to help them tell the story. The children practice tapping their sticks in different ways before the story begins. Shelby comments, "Many of you are tapping loudly. There will be part of the story when you can do that. There is also a part when you will tap softly. Can you show me how that would sound?"*
>
> *Shelby begins, "Once upon a time, there were three billy goats that were so hungry they decided to go up a big hill to a certain spot where the best, greenest, juiciest grass*

118

grew. They had to cross a bridge over a big river to get to this special spot and an ugly, mean troll lived under the bridge just waiting for a tender, tasty goat to walk across his bridge so he could gobble him up." She uses her hands to demonstrate in the air the direction words *over, under,* and *across. "Now these three goats were brothers,"* Shelby continues, *"and they were all named Gruff. Big Billy Goat Gruff was the oldest and biggest, Middle Billy Goat Gruff was not quite as big or as old as his Big Billy Goat Gruff brother, and, finally, Little Billy Goat Gruff was the youngest and the smallest of the three."* Shelby holds up her fingers to indicate one, two, three. *"Since Little Billy Goat Gruff was the youngest, his brothers made him cross the bridge first,"* Shelby reads.

At this point Shelby pauses and Carol says to the children, *"I wonder what kind of sound Little Billy Goat makes when he crosses a bridge."* Hasna taps her rhythm sticks softly and says, *"Trip-trap, trip-trap."* Carol says in a very soft (and "tiny" voice), *"Hasna thinks he would sound like this,"* and she imitates Hasna's soft tapping and the words trip-trap-trip-trap-trip-trap. *"The words trip-trap-trip-trap-trip-trap make a pattern,"* she says. Eva adds, *"He's the baby — that's why he has to go first; he is so teeny-tiny that his feet aren't big enough to make a big noise"* (KDIs 31, 36).

Shelby continues with the story: *"Little Billy starts across the bridge. He was so little that his feet made a little sound across the bridge."* The children join her, quietly tapping with their rhythm sticks, as Shelby whispers, *"Trip-trap, trip-trap. Suddenly, the troll speaks up in his ugly, mean troll voice...."* She pauses and says, *"What kind of a voice does a troll have?"* The children make loud, grrrrr sounds. *"Oh,"* says Shelby, *"troll voices are very loud!"* Cody adds, *"Louder than a thousand lions!"* (KDI 36). Shelby continues in a loud, growly troll voice, *"Who's that tripping over my bridge? I am going to come up there and gobble you up."* And then, using her regular voice, Shelby continues, *"It is I, Little Billy, the tiniest billy goat."* Shelby crouches to make her body smaller like the little goat. Bella interrupts, *"Little goats don't talk like that. They sound higher and softer"* (KDI 36), she says, making the Little Billy voice for them. *"Oh,"* says Carol, *"let's all make little voices like the littlest billy goat."* They all repeat the words in higher, softer voices, *"It's me, Little Billy, the tiniest billy goat"* (KDI 36). Shelby continues, *"Wait for my brother, Middle Billy. He's coming next and he is much bigger!"*

Support strategies

- As Shelby is describing the story for the children, she uses a number of spatial relations words: "Today we are going to tell you a story about some hungry goats that decide to *go up* the hill so they could fill up on their favorite grass, which only grows *in* a special place *on* the hill." She uses her hands to emphasize the direction words and to call the children's attention to them.

- As the story begins, Shelby reads more spatial terms, such as "*over* a big river" and "*under* the bridge." She uses more hand motions to help children understand the position words.

- When Shelby begins to describe each of the billy goats, she uses several measurement terms: Big Billy Goat Gruff was the *oldest* and *biggest,* Middle Billy Goat Gruff was *not quite as big* or *as old as* his Big

While reading The Three Billy Goats Gruff, *the teachers give children rhythm sticks, which they are encouraged to use to tap softly or loudly in accordance with the story.*

Billy Goat Gruff brother, and, finally, Little Billy Goat Gruff was the *youngest* and *the smallest* of the three.

- Shelby holds up and counts off three fingers to emphasize that there are three goats. In relating the story, she also says the littlest goat crosses the bridge *first*. Thus, she introduces cardinal and ordinal number words in the context of the narrative.

- When Carol hears Hasna's response (Hasna taps her rhythm sticks softly and says "Trip-trap, trip-trap"), Carol repeats it to make sure that the rest of the class can hear it and points out that the words form a pattern.

- When it is time to establish the sound of the troll's voice, the children are quite specific about it, growling all together. Shelby describes the sound they make: "Troll voices are very loud."

- Shelby crouches down trying to be small but uses her regular voice to talk like Little Billy and is interrupted by a child who tells her that "Little goats....sound higher and softer."

- Shelby stops for a moment, and Carol invites the children to express their own understanding of "a little" sound by asking them to make the kind of sound they think the littlest of the brothers would make when he crosses the bridge.

- Shelby continues the story, including the number word and the measuring terms: *tiniest, little, middle,* and *bigger.*

Math in Action: Let's Have a Parade!

Carol and Shelby have planned this large-group time primarily with the movement KDIs in mind (KDI 16. Gross-motor skills, KDI 17. Fine-motor skills, and KDI 18. Body awareness). They anticipate that, like many other movement activities, there will be a number of opportunities for children to develop spatial awareness as they move their bodies and the scarves they hold in their hands. In describing the parades they have seen, the children also introduce other mathematical concepts that the teachers repeat and expand on.

> As the last few children finish their snack and join large-group time, Carol says, "I heard many of you talking about going to see the parade over the weekend. I'd like to hear what you learned about parades." The children share their observations. Jamal says, "There are marchers and fire trucks." "The police cars are first" (KDI 31), Xavier adds. "Oh," says Carol, "the police cars are first *and* the marchers are second?" Bella adds, "And they have lots of flags" (KDI 32). "You get lots of candy," Luis says. "The lady in the back of the truck throws it," says Margie, "and I caught three" (KDIs 32, 35).
>
> Carol continues, "Some of you said that sometime you would like to have your own parade. And that's what we thought we would do today for large-group time. Shelby has scarves for each of you to wave, sort of like flags, and I'm going to put on some music so we can march around the room." Carol motions around the room with her hand. "When it's your turn to lead the parade, you can show us how to wave our scarves," explains Carol. "I want to be first" (KDI 31), says Rene. The children crowd around Shelby. "Can I have two scarves," asks Margie, "one for each hand?" (KDI 32). "You want two scarves for two hands, Margie?" asks Shelby. "This time, we have just enough scarves for each child to have one," Shelby explains. "We can get more scarves another time and do a scarf in each hand."
>
> Once all the children have a scarf, Shelby says, "Rene asked to be the first leader. We can all march behind her and wave our scarves over our heads the way she does." Carol turns on the music and Rene begins to march, waving her scarf over her head. Shelby asks Rene to describe where she is going to lead the parade. Rene announces, "First we go around the block area and then through the house area" (KDIs 31, 35).

Scarves offer lots of opportunities for exploring spatial awareness terms, including in front of *and* over.

Support strategies

- Carol repeats Xavier's words, "The police cars are *first* and the marchers are *second*?" to emphasize and extend his use of ordinal words.

- Shelby supports Margie's understanding of number by restating "You want *two* scarves for *two* hands...." and then explains that in order for everyone to have two, they would need more scarves.

- As they march, Shelby suggests that Rene describe where she will be leading the parade, knowing that it might encourage her to use some position and direction words.

Math in Action: Five in the Bed[1]

At today's large-group time, Jackson chooses the fingerplay "Ten in the Bed" from the classroom song book. Carol and Shelby think the children are ready to act it out since they are already familiar with the song. It will give the children another concrete way to practice counting and to understand the concept of *take away* (subtraction).

> *Since this is the first time the children are acting out the fingerplay, the teachers begin with 5 (rather than 10) children in the bed. "Here is the bed," says Carol, as she points to the blanket. "Let's start with five children sleeping in the bed."*
>
> *Carol names three children who lie down on the blanket and says, "Let's see. We need five and we have one, two, three. How many more do we need to have five on the bed?" Lian says, "Two more, I'll do it!"* (KDI 32). *Carol says, "Two more.*

> *Okay, then Lian is the fourth and Blake makes five. Now we have five children on the bed." Carol and Shelby begin the song: "There were five in the bed," and the children join in, "And the little one said, 'Roll over, roll over.' So they all rolled over and one fell out"* (KDIs 32, 35). *Shelby says, "Austin, now that you fell out of bed, I wonder how many are left in the bed?" "Let's count together," suggests Carol. All the children begin counting at once and Carol responds, "Oh my, you're all counting and saying numbers and I'm having a hard time keeping track. Let's count together as we point to each child left." They count, saying (all at once) "One, two, three, four," as they point to each child left in the bed. "There are four kids left!"* (KDI 32), *they shout. Shelby nods her head and says, "There are four. We started with five, Austin fell out, so now we have four children left." They begin the song again and continue to repeat it until there are no children left "in the bed."*

Support strategies

- Carol models counting and poses a math question, "How many more do we need to have five on the bed?" She acknowledges Lian's answer.

- Carol mixes an ordinal and a cardinal number when she says, "Lian is the fourth and Blake makes five."

- After Austin rolls off the blanket, Shelby asks how many children are left in the bed.

- To determine how many children are remaining, Carol invites the children to count together as they simultaneously point to each child left on the blanket (bed).

---

[1]Adapted from activity 43 in *50 Large-Group Activities for Active Learners* (Boisvert & Gainsley, 2006, pp. 104–105).

Large-Group Time

While singing "Ten in the Bed" during this large-group time, the teachers model using their fingers to help children count.

- Shelby supports the idea of cardinality by emphasizing that the last number counted tells how many children are left on the blanket, "There are four. We started with five, Austin fell out, so now we have four children left."

- Since children need repetition to understand math concepts, the teachers repeat "one" child falling out of the bed until there are no children remaining on the blanket.

10

Cleanup Time, Mealtimes, and Transitions

Clean up your own mess.
— Fulghum (1990, p. 6)

Cleanup Time

Teachers frequently view cleanup time as one of the most stressful parts of the day — as a chore to be endured, rather than as a learning experience, like any other part of the daily routine. Children, even if they don't feel this way to begin with, may pick up on this negative attitude. In HighScope classrooms, adults plan daily strategies and experiences for cleanup time just as they plan for large-group time or any other component of the daily routine (Evans, 2007). Thinking this way can help adults, as well as children, approach cleanup time in a more positive manner.

What Children and Adults Do During Cleanup Time

The purpose of cleanup time should not be to restore the room to a pristine condition (real "cleaning" for safety and sanitation is done by adults after the children leave). Rather, cleanup time helps children understand how the room is organized (the functional or physical reason things go in a particular area and/or are grouped together) so they can act independently in the find-use-return cycle of play (see the sidebar on p. 127 for a detailed explanation of the find-use-return cycle). It also provides an environment for the children and adults in the classroom to work on building a sense of community by encouraging shared control, collaboration, and responsibility. If adults do not intentionally plan for cleanup time, thinking about possible activities or strategies, then it is likely to deteriorate into chaos and conflict. Here are five simple suggestions that HighScope teachers keep in mind when planning for cleanup time:

1. Since young children yearn to do things independently and also like to do things they can be successful at, organize cleanup to promote children's independence and feelings of competence.

2. Plan ways to help children disengage from activities they are deeply involved in at the end of work time.

Meaningful Math in Preschool

To help prevent children from becoming frustrated if they are still working on a project when it is cleanup time, children use work-in-progress signs over their unfinished projects so they can continue to work on them the next day.

3. When children have spent all of work time on a project or event and they have no time to actually follow through on their plan when work time ends, they are bound to be frustrated. In the HighScope classroom, we use work-in-progress signs that children can put over unfinished work to be saved.

4. Break down large cleanup tasks into smaller jobs.

5. Use group discussion when cleanup fizzles. (Gainsley, 2005, pp. 2–3)

Cleanup time also offers a variety of learning opportunities. As you will see in the following anecdotes and the longer scenario that follows, math learning can take place during cleanup time just as easily as it does during any other part of the day. Some children may learn more about shapes (KDI 34) and spatial awareness (KDI 35) as they locate and put away the different-shaped unit blocks. Others are challenged to see how many items they can return to their correct area (KDI 32). Still others may be putting away some of the cardboard tubes and comment that one of them is longer than the other (KDI 36). A child may note that more children are putting things away in the block area than other areas and connect this observation to the fact that more children played in the block area (KDI 39). It is busy and sometimes noisy during cleanup time, so be sure to tune in your math senses so you will be able to catch the math learning that is taking place.

The Find-Use-Return Cycle

The daily routine and, in particular, cleanup time, goes more smoothly if the learning environment is set up with the find-use-return cycle in mind. Accessible storage of materials is the critical factor in making it possible for children to develop independence by finding and returning the materials they need. Following are some useful guidelines to follow, if you are trying to support the find-use-return cycle:

- **Store similar things together.** Go beyond simply putting art supplies in the art area: Put all the drawing tools (e.g., markers, crayons, chalk, colored pencils) on one shelf and all the paper in one area (e.g., plain white paper, construction paper, wrapping paper, miscellaneous paper scraps, tin foil, wax paper). Putting materials that share similar functions close together not only facilitates cleanup but also gives children ideas for alternative materials.

- **Use containers children can see into and handle.** Children need to be able to find the materials they want to play with. Use open, easy-to-handle containers and tubs that can be stored on low, open shelves. Open, low, flat containers are good for holding smaller materials while still remaining very visible.

- **Label containers in a way that makes sense to children.** Label both the container and the shelf (or floor) where the material is stored. This provides a fixed storage space that both teachers and children can rely on to locate the material they need. It is very important that children understand the labels. These may include labels made from the actual material, tracings, drawings, catalog pictures, photos, photocopies of the material, or any of these along with the word of the material.

Over There! Catch the Math During Cleanup Time

KDI 31. Number words and symbols
KDI 33. Part-whole relationships

Theresa and Clay (the teachers) distribute "cleanup tickets" (number cards). Faith says, "I want a number card with a 2 on it." Theresa responds, "Sorry, I don't have any 2s." Faith says, "Or you could just use a 1 and a 1 and that's 2."

KDI 32. Counting

Clay approaches Willem, who is sitting in the book area, and says, "Let's find two more things to put away." Willem picks up two Matchbox cars and puts them back in their tub.

KDI 31. Number words and symbols
KDI 32. Counting

Theresa asks Rowan, "Would you go to the house area and find three things to put away?" Rowan picks up a cookie cutter, a fork, and a bowl and says, "See? I have three things; I'll take the cookie cutter to the art area first."

KDI 34. Shapes

Kiefer is cleaning up the Magna-Tiles when Clay (a teacher) comes by. Kiefer says, "I'm putting them away by shape, first the big squares, then the small squares, then the long pointy triangles."

KDI 34. Shapes

Natalia and Cora are in the house area, cleaning up the poker chips. Cora says, "These are all round; do you think there are any square ones?"

KDI 35. Spatial awareness

As she is putting away a puzzle, Faith finds a piece she was looking for and says, "This piece was hiding under that one."

KDI 36. Measuring

Brynna is upset when cleanup time begins, following the five-minute warning. Clay (a teacher) says, "You look frustrated. I can see that your face looks upset." Brynna clarifies, "No, this is my sad face because it only took one minute!" (The five-minute transition from work time to cleanup time felt more like one minute to Brynna.)

KDI 36. Measuring

During cleanup time, some of the children "ride" the big hollow blocks to the shelf. Curtis sits on the back of Desmond's block, and Desmond says, "You are much too heavy."

KDI 37. Unit

While she is cleaning up her snack, Faith lines up the leftover fish crackers along one edge of her napkin, counts them, and says, "My napkin is seven fish crackers and a little fish nose long."

KDI 38. Patterns

Camille and Franny take turns tossing Duplos into the Duplo tub. Franny explains, "We're doing a cleanup pattern. Watch! First Camille throws in a red one, then I throw in a green one, then Camille tosses in another red one and I toss in a green one." They keep up this pattern until they put away all the green and red Duplos.

Math in Action: Cleanup Tickets

As a cleanup strategy, teachers Theresa and Clay are passing out the cleanup tickets they prepared in advance. Each ticket is made from an index card that has a numeral from one to five written on one side and the same number of dots on the other side. The teachers think that this strategy will help make cleanup more interesting and also support children's math learning. Let's take a look and see what happens.

As Theresa hands out the tickets to the children, she says, "Look at the number on your ticket on one side or count the number of dots you see on the other side to find out how many things you should put away. When you put away the number of things on your ticket, you can bring it back for a new one." As she hands out the tickets, she thinks about children's counting ability and gives the larger numbers to the more practiced counters and experienced "cleaner-uppers."

Theresa holds up her own ticket and says, "I have the number 3 on my ticket and one, two, three — three — dots on the other side. I'm going to put away three things." She reaches down and counts out loud, "One, two, three," as she picks up three blocks and puts them in their places on the shelf. "Now I need another ticket. I wonder what number I'll get next," she says.

Clay approaches Ellie and hands her a ticket. "Here's your cleanup ticket, Ellie. Let's see how many things you are going

Cleanup time affords many math learning experiences. Here a teacher and child count the number of beanbags as the child places them in a bag.

to put away." He points to the numeral 2 on the ticket and says, "That's the number two. Let's find two things to pick up." Ellie does not indicate that she is going to pick anything up so Clay picks up a baby bottle and bib that are on the floor. He says to Ellie, "One, two. Here are two things for you to put away." He puts them in Ellie's hand and walks with her to the house area to put the bottle and bib away. "Now we can get another ticket," he says.

Devin looks at his ticket and says, "Five, that's too much. I want three" (KDIs 31, 32). He looks through the pile of extra tickets on the table and finds the numeral 3 (KDI 31). Theresa watches and says, "Three is *less than* five. After you find three things, will you look for a number *more* or *less than* three?" "Another little number" (KDI 31), replies Devin.

Jordan comments, "I'm putting away lots of things today" (KDI 32). Clay responds to him, "Lots?" "Yep," says Jordan. "First I put away three things. Then I put away four things" (KDI 32). "Yep, that's a lot of things," acknowledges Clay. "Three things plus four things." "Seven things!" (KDI 33) exclaims Jordan. "I put away two things plus five things," says Clay. "That's seven too" (KDI 33), says Jordan. "Hey," says Clay, "Three plus four is seven and two plus five is also seven." "Yeah," says Jordan.

Support strategies

- Clay and Theresa use a cleanup strategy that supports counting, part-whole recognition, and numeral recognition.

- Theresa chooses which tickets to give children based on their counting ability to ensure their success.

Meaningful Math in Preschool

Possibilities for math talk abound during mealtime. "Look, I poured just a little milk without spilling," says this child.

- Theresa models counting the dots on her ticket and makes the connection between quantity and a number symbol: "I have the number 3 on my ticket and one, two, three — three — dots on the other side."

- Clay names the numeral on Ellie's ticket and models counting as he picks up two things for Ellie to put away.

- Theresa acknowledges Devin's understanding of quantity: "Three *is* less than five. After you find three things, will you look for a number *more* or *less* than three?"

- Clay encourages Jordan to explain what he means when he says that he picked up "lots" of things. When Jordan uses addition to figure out the total number of things he has picked up, Clay adds his own addition fact. He then points out that there are different combinations of numbers whose sum is seven.

Mealtimes

In most part-day preschool programs, children and adults sit down together either to share a snack provided by the program or eat one brought from home, while, in full-day programs, they may gather for one or two meals. Since the focus during these parts of the day is on social interaction, it is preferable for children to eat with the same adult and children who gather together for planning, recall, and small-group times (Marshall, 2007).

130

What Children and Adults Do During Mealtimes

In addition to refueling them for what lays ahead, meal- and snacktimes provide children with a familiar social setting that invites casual talk, one-on-one conversations, or table-wide discussions. It is important for adults to sit down and eat with children, not just as a natural social situation but also as an opportunity to share relaxed conversation and share children's ideas (Marshall, 2007). Note the emphasis on "relaxed" conversation — this is not the time to bring out a special "snacktime" lesson. Countless (mutually beneficial) opportunities for teaching and learning naturally arise during these times of the day (see the sidebar on p. 132 for ideas for math talk at mealtimes). Why don't we listen in on some of these conversations and see if we catch any math learning taking place?

Over There! Catch the Math During Mealtimes

KDI 31. Number words and symbols

During lunch, Ella, Jayla, and Brianna chat with Becky (a teacher) about birthdays. Ella discovers that Jayla and Brianna also have birthdays this month. They compare dates. Ella says, "Hey! I'm five first!"

KDI 31. Number words and symbols
KDI 32. Counting

Landon says, "I'm three, so I get three of these," and he takes three crackers.

KDI 32. Counting

Nygel counted his 14 crackers with the help of Judah (a teacher), who points to each one as Nygel counts. When Nygel has finished eating most of them, he observes that there are "only five left," and pops that one in his mouth. Nygel notes that "there are four left," and this process continues until Nygel announces that "I ate all the crackers!"

KDI 34. Shapes

David holds up a cheese cracker and says, "It's a diamond or a square. I see a circle on it too."

KDI 34. Shapes

While reading the shape circus book during lunch, Franny talks about a triangle she spots and says, "It has three sides."

KDI 35. Spatial awareness

During lunch, Brynna, Kale, and Gabriel talk about moving into new apartments. Kale adds, "Ella moved. She lives next to us now."

KDI 35. Spatial awareness

While eating an apple, Gabriel explains to Marlee, "Apples come from apples and sauce comes from a little village way over there, far, far away."

KDI 31. Number words and symbols
KDI 36. Measuring

After Kiefer points out that there are two numeral 2s on the milk carton, Claire says, "Yeah, but that 2 is bigger than that 2 (pointing to the other one); that one's smaller."

KDI 36. Measuring

At breakfast, Landon says, "I have more milk left than she does."

KDI 37. Unit

During lunch, Hayden says to Carson, "Look, my napkin is 18 raisins wide!"

KDI 38. Patterns

During lunch, Jayla begins to make a pattern using her cucumber slices and carrot sticks: green, orange; green, orange; green, orange. Claire picks one cucumber slice and one carrot stick from her plate and adds them in order to continue Jayla's pattern.

KDI 38. Patterns

Josh lines up his snack mix like this: two raisins, one pretzel; two raisins, one pretzel; two raisins, one pretzel. "See Judah," he says, "I made a pattern with some of my snack; there aren't any Cheerios, because I ate them!"

KDI 39. Data analysis

During Friday's snack, the children decide how many bags of certain ingredients their teacher Becky will buy for next week's trail mix. Becky has a chart ready with pictures of each ingredient at the top of a column: pretzels, raisins, sesame sticks, and cheese crackers. The children put a check mark in the columns of the foods they like to have in the mix. When the chart is complete, Becky asks, "What should I put on my grocery list?" "Everybody likes pretzels," says Gina. "Lots of kids like raisins," observes Malcolm. "Pretty many checks next to the cheese crackers," notes Zak. "What about the sesame sticks?" asks Becky. "No checks there. That means nobody likes them," says Noah.

"Math Talk" at Meals

Here are some ways to encourage math talk at meals:

- When children are eating, try using language that describes the position of things, for example: "You put your raisins on top of the peanut butter." "Your cup is *between* the spoon and the plate." "Gabby had to go *under* the table to get her napkin."

- Children like to talk about their ages. Encourage them to talk about how old they are, who is younger or older, and the ages of their siblings (if they know). Give them time to think about it. For example:

Adult: Shannon, you said you are four, but your baby is just a baby. (The adult pauses.)

Shannon: Yes, he's just little (KDI 35). (The adult listens and pauses.)

Micah: My baby's bigger now. She's two (KDIs 31, 36).

Adult: Two is getting bigger. (The adult pauses.)

Kobe: Yeah, but not as big as me — I'm three! (KDIs 31, 36).

Source: Adapted from Marshall (2007, pp. 58, 65, 91, 114, 138).

Math in Action: Cheese and Crackers for Snack

Lamonte lines up his crackers on his napkin. Then he puts one of the cheese cubes on top of each cracker. He says to Janeen (a teacher), "I have four crackers and four cheeses, and then two more crackers" (KDI 32). She responds, "You put them in a line so you have one cube of cheese on top of each round cracker. Do you want anything more?" She pauses. Lamonte thinks for a minute and says, "Two cheeses; then there'll be a cheese for each cracker" (KDI 32).

A child across from Lamonte says, "I like flat cheese better than cheese cubes" (KDI 34). Another child adds, "I like cheese sticks because they are long and you can pull them apart" (KDI 35). Janeen comments, "Food comes in so many different shapes. Cheese cubes, circle crackers. Can you think of other kinds of food that come in shapes?" "Pancakes come in circles" (KDI 34), says Lamonte. Lousia laughs and says, "Fish crackers are shaped like fish."

Support strategies

- Janeen acknowledges Lamonte's understanding of one-to-one correspondence. She not only names the shapes of the cheese and cracker but also adds a spatial term when she restates his description: "You put them in a *line* so you have one cube of cheese on top of each round cracker."

During snacktime, preschoolers often encounter experiences with one-to-one correspondence as these children do: Each child gets one muffin.

- Curious to see if he is interested in adding cheese to his remaining crackers, Janeen asks Lamonte, "Do you want anything more?" and pauses to give him time to decide.
- When the other children begin to talk about attributes of cheese they like (e.g., "they are long and you can pull them apart"), Janeen offers a gentle extension when she says, "Food comes in so many different shapes. Cheese cubes, circle crackers. Can you think of other kinds of food that come in shapes?"

Transitions

Transitions can be a change of activity, place, or the people one interacts with. Children in preschool move through many transitions in a day — leaving home for preschool, saying good-bye to mommy or daddy, moving from snacktime to large-group time, going from outside time to naptime, leaving school to go home, or changing classrooms or locations. Some children handle transitions well, seeing an opportunity to move on to something new and exciting, while others find transitions disruptive. By making transitions few in number, short, and physically and mentally engaging, adults can accommodate children who have different reactions to change.

What Children and Adults Do During Transitions

Children are born with innate temperamental differences that affect how they react to change and transition. For some children, it is a smooth process. For others, it produces uncertainty and anxiety. When children do not understand why they have to stop what they are doing and move, "their confusion and frustration can lead to behavior upsets" (Evans, 2007, p. 1). Adults play a critical role in helping both the most easygoing and the most anxious children handle transitions well and enabling them to feel a sense of control when faced with new activities, people, or situations.

We know that active learning and support from adults are critical components in children's early learning; keep the need for these in mind as you plan for the transitions (both expected and unexpected) throughout your day. The necessity of establishing and maintaining a consistent daily routine helps children know what is going to come next. At the beginning of the year, as you think about the daily routine, it is very important to try to keep the number of transitions between activities, places, and people to a minimum (Epstein & Hohmann, 2012). Some fundamentals for ensuring that transitions go smoothly include the following strategies:

- Begin new activities promptly. Children get "antsy" when they are forced to wait for the whole group to gather. Plan a "beginning" component of the activity that occupies children.
- Plan ways to avoid waiting times. If children must wait, be prepared with ideas for keeping them actively involved.
- Think of fun ways for a group of children to move from place to place (around the program).

See the sidebar on the facing page for additional strategies.

When you apply these fundamentals, transitions are not only fun but also present opportunities for learning in all content areas, including math, as you will recognize in the anecdotes on the following pages.

Transition Strategies

Like so many other strategies teachers use throughout the day, there is no "one-size-fits-all" approach that works for all children. Following the next four fundamental transition strategies consistently will help adults guide children successfully through a wide variety of transitions.

- **Take the child's perspective, keeping the developmental characteristics of preschoolers in mind.** Children at this level of development are usually able to keep only one or two things in mind at a time, so they can easily become unfocused and forget what they are supposed to do next. Thus, it is essential to be patient and willing to consider children's needs along with your own. Adults can best help children predict the beginning of a transition by not depending too heavily on conventional time units and by not referring to time units at all if the transition is more than 10 minutes away.

- **As you plan what to do during shifts in activities, consider what distractions or additional tasks you will have.** If tables must be cleaned off as children move from table activities to hand washing, if cots must be laid out as children arrive from outside, or if parents are trying to talk to you as you help children brush their teeth after breakfast, it will be difficult to give children the support they need. If there are tasks to be done during transitions, make a clear plan for who will do what. If this is not possible, using a well-known song or a familiar activity (plan this beforehand) may help children manage the transition more independently.

- **For each shift in activities, make a plan that includes an activity with choices, a warning, and extra time for the transition.** Success during transitions is all about predictability and a sense of control, with a good dash of fun included.

- **Give children the time they need to move through transitions playfully and to express any feelings they have.** As the transition begins, be ready to acknowledge children's feelings. Children may be excited about, or resistant, to the change — give all such emotions time for full expression, rather than trying to suppress or stop them. It is always best to take an unhurried approach to transitions, moving through the transition both playfully and slowly so that children enjoy the process and have time to develop the skills to negotiate other transitions successfully.

Source: Adapted from Evans (2007, pp. 3–5).

Over There! Catch the Math During Transitions

Math language during transitions comes from two sources. First, teachers intentionally plan transitions that incorporate math, for example, using an attribute — such as the number of buttons or the pattern on children's clothing — as the basis for determining the order in which they move to the next activity. Second, math language and thinking can also come from the children themselves, who may become aware of sequencing, spatial characteristics, or other mathematical properties as they move through the day.

KDI 31. Number words and symbols

Mariel is getting ready to go outside and looks at the clothes chart. She says out loud to herself, "First come the snow pants. Second are the boots. Third is my jacket…"

KDI 31. Number words and symbols

Eddy goes to the sink while saying "I'm first to wash my hands!"

KDI 32. Counting

After greeting time, Erica (the teacher) says, "All the kids at Christine's table, take five giant steps." The children count as they take each step.

KDI 34. Shapes

At the end of recall time, Christine says, "Anyone with circles on their shirt can go wash their hands for snack." Noah and Abe get up to go to the sink to wash their hands.

KDI 35. Spatial awareness

Struggling to put on her jacket, Brynna goes to Marcella (a teacher) and asks, "Can you help me? I can't get my hands in the sleeves. They're outside out."

KDI 35. Spatial awareness

The classroom has a map (located at the children's eye level) of where the cots go. Each cot on the map is labeled with each child's name and symbol. Tomas looks at cot map and says, "I'm between Bonita and Fernando."

KDI 36. Measuring

After small-group time, Poppy is washing glue off her hands at the sink and says, "This dried glue takes longer to get off than the paint does."

KDI 36. Measuring

Peter and Lev are getting their boots on and Peter says, "Look at our feet! My feet are way bigger than yours."

KDI 37. Unit

Zanni uses the measuring tape to see how tall the biggest sand timer is. "It's 17," she says.

KDI 38. Patterns

"I'm going to do a walk-jump-walk-jump-walk-jump pattern to my cubby," says Adam.

KDI 38. Patterns

At the end of large-group time, Tara (a teacher) says, "If you can find a pattern on your clothes or someone else's, you can go get your coat on for outside time." George says, "You have blue–white-blue-white-blue-white-blue-white stripes on your shirt!"

Math in Action: Going to the Lunchroom

Campbell and Chantal's class are on their way to the lunchroom. Chantal asks Holly, "How should we move first today?" Holly says, "Three jumps" (KDI 32). Chantal continues, "Okay, first we will jump three times. What should we do second, Drew?" Drew responds, "Take three giant steps" (KDIs 32, 35). Campbell says, "Let's count three jumps together." All the children count, "One, two, three" (KDI 32) as they jump. "Now let's count three giant steps," Campbell says. And they count as they take giant steps toward the door, saying "One, two, three" (KDI 32).

As they are on their way out the door, Chantal says, "So what's next?" The children say, "Three jumps" and down the hall they go, alternating three giant steps with three jumps until they finally reach the lunchroom. When they arrive, Holly says excitedly to Chantal, "You know what?" "What?" responds Chantal. "We made a pattern all the way here" (KDI 38)! Holly exclaims. "How did we make the pattern?" asks Chantal. Holly grins and says, "We did three jumps, three giant steps; three jumps, three giant steps; three jumps, three giant steps" (KDI 38)!

Support strategies

- Chantal models the use of ordinal number words (*first* and *second*) when she asks the children how they will move.
- Before they get out into the hall, the children count the three jumps and three giant steps together, so they experience one-to-one correspondence as they simultaneously say the number and do the movement (i.e., they say "one" as they jump the first jump).
- When Holly says, "We made a pattern all the way here!" Chantal is curious to find out whether Holly understands how the pattern was formed and offers Holly a gentle extension by asking "How did we make the pattern?"

Getting ready to go outside gives children lots of opportunities to develop their spatial awareness skills: how to put on shoes on the right feet and how to help a friend zipper a sweater, even though it is on backwards!

What Happens When Children Have to Wait?

In her book *"I Know What's Next!" Preschool Transitions Without Tears or Turmoil*, Betsy Evans (2007) aptly answers this question saying, "Expecting young children to wait with nothing to do is an invitation to chaos" (p. 67). Whether they are sitting at a table while you get materials ready for small-group time or waiting until it is time to take them to the gym, you are setting them up for failure. While lining up is an age-old tradition, beginning in preschool, today it is considered by many to be not only a waste of time for both the children and adults but developmentally inappropriate as well.

Throughout this book we have stressed that children are active learners "who learn best by using materials, making choices, moving their hands or whole bodies, singing or talking. They can be learning all the time, as long as they as active with their minds or bodies" (p. 66). Most important, review your daily routine and look for ways to completely eliminate waiting times. If that is not possible, Evans recommends "circling up" rather than lining up, since children in lines are more likely to incite spatial and physical disturbances. Children are accustomed to doing things in a circle, and this formation more easily adapts to different situations.

The secret to the success of these and the other in-class transitions is thorough planning. As your teaching team plans for the day, be sure to plan for each transition too, keeping in mind the following considerations: how long the transition will be, whether you will be moving to a different room or location, or if you will be going outside. Following are some ideas for making the experience of waiting and moving from room to room, fun and math rich:

- **I spy — Recognizing numerals.** "I spy with my little eye, the numeral that comes after 2. What could it be?"
- **I spy — Recognizing things to count.** "I spy with my little eye, pots hanging on the wall. How many pots in all?"
- **I spy — Recognizing shapes.** "I spy with my little eye, something wooden with three points. Which shape do you think it is?"
- **I spy — Recognizing spatial relationships among people and objects.** "I spy with my little eye, something under Sue's table that they used during small-group time."[1]
- **A song for recognizing spatial relationships among people and objects.** To the tune of "If You're Happy and You Know It," insert a line that would make use of the children's spatial awareness, for example, "If you're happy and you know it, put your hands above your head. Okay, Noelle, where should we put our hands next?" Noelle might reply, "Under our arms" (KDI 35).
- **An obstacle for recognizing spatial relationships among people and objects.** If you are changing location (e.g., leaving the classroom), put a hoop, block, or small slide in the doorway (sing a song that gives you the delay you need to set this up), and then ask children how they're going to get past the obstacle (i.e., *through, over,* or *around* the obstacle). For example,

Teacher: Hey, little duckie by my side (the next child in line). How will you get through the door?

Child: I'm going over the block (KDI 35).

The "I spy" ideas are adapted from Evans (2007, pp. 70–73, 79).

11
Outside Time

Nature provides almost limitless opportunities for hands-on learning. Children can turn over a rock or a piece of wood and find insects, worms, plant roots, and fungi. They can look at clouds or collect leaves and gain an appreciation of the variety of forms and textures in nature. They can hear birds or feel the wind and become aware of a whole sensory world beyond the classroom door. They feel joy.

— Handler & Epstein (2010, pp. 1–2)

During outside time, children not only exercise their large muscles but also observe, interact, explore, count, measure, estimate, and much more. This is a time for running and riding, for balancing and building, for digging and daring, for exploring nature, and for adventurous as well as quiet role play. The playground is a "sanctuary for types of play that may not be permitted in the classroom, center, or home" (Harris, 1996, p. 120).

What Children and Adults Do During Outside Time

Whoosh! The door opens and the preschool class pours out: Several children run up and then down the hill; three go over to get a turn on the swings; one goes back inside to get the biggest sand timer; some get the pots and pans and other utensils out of the shed and move across to the sandbox; two are waiting for an adult so they can hammer nails into a stump; and still others are getting out the "big wheels," telling each other how fast they will go today! While the children embrace all the adventures that the outdoors affords them, the adults are active participants in their play, playing catch, pitching balls, examining a pattern on a rock, pushing a child on the swing, helping to prepare cookies in the sandbox, or watching children blow bubbles.

If you look carefully, you'll see that math plays an important role in children's outside play, including lots of measuring opportunities in the sandbox.

During outside time, adults are active participants in the children's play.

Outside time is not break time for the adults, nor should teachers use this time as a period to chat with each other or with other adults. Although they may be monitoring for safety, sharing observations with other teachers, or connecting with parents, teachers need to remember that supporting the interests of children, responding to their questions, going to see the "deepest hole ever," and being able to build on these learning opportunities is their first and most important responsibility.

Over There! Catch the Math During Outside Time

Learning doesn't stop when children go through the door to the outside environment. The outdoors is not just a place for children to use "outside voices," release pent-up energy, or make rivers in the sandbox. It is a total learning environment — an open-air classroom annex where they can explore many of the same concepts (though perhaps at a higher noise level or at a greater speed) they explore inside. The nine mathematics KDIs turn up and play an important role in children's outdoor play too. Let's see how sharp our math senses are as we observe a group of young children at outside time.

KDI 31. Number words and symbols

Chantal, Abby, and Quinn get ready to run around the climber. Abby puts sticky notes with numbers on each child and says, "Chantal, you're 1; Quinn, you're 2; and I'm 3."

KDI 31. Number words and symbols

As the class arrives at the park, Tamsin and Eddy run toward the climbing wall. "I'm first," says Tamsin. "Okay, I'm second," sighs Eddy.

KDI 32. Counting

"Hey, Carl," says Tommy. "Here are four more of the speckled, orange rocks we found yesterday. See? One, two, three, four."

KDI 32. Counting

Gelina lines up all the dinosaurs on the wall, one behind the other, and then counts them: "1, 2, 3, 5, 6, 7, 9, 10. I have lots of dinosaurs!"

The opportunities for counting things outside, as noted above, are plentiful. Initially, however, you may support children in identifying places for spotting numerals (e.g., mailboxes, license plates, house numbers), especially if you are in suburban or rural settings. The outdoors is also a rich source of shapes, natural and manmade, as illustrated in these three anecdotes below.

KDI 34. Shapes

"Hey, Linda!" calls Brynna. "The sun is so bright I put on the round sunglasses!"

KDI 34. Shapes

On their way to the park, Kobe looks at the building across the street and says, "Look at that weird window over there. It's a triangle!"

KDI 35. Spatial awareness

Hayley and Jordyn watch two squirrels run around the playground. Hayley says, "Look one is going up the tree!" "And the other one is right behind him," adds Jordyn.

To help support chidren's developing spatial awareness, take turns playing I spy when walking to a park for outside time, using both shapes and different position and direction words. For example, you might say, "I hear the subway *under* the sidewalk" or, while passing a no-parking sign, "I spy a rectangle sign *above* your head."

If you live in snow country, you know that children often notice how deep the snow is, for example, you might hear, "The snow covered the bottom of the slide" or "It's up to my knees!" Some children may notice that the snow depth changes from day to day or they may be interested in using a yardstick or measuring tape to discover exactly how deep it is.

If the children are curious about how much rain falls, they can set a clear plastic cup out on a window sill, porch, or fire escape and when the rain stops, bring it in and guess or measure it in a measuring cup. They might also choose to draw their own measuring lines on the cup.

Finally, a sturdy measuring tape is a useful tool to have on hand and keep with your other outdoor materials. There is always something to measure outside — from the length of a worm to the height of the climber as shown in the following examples.

KDI 36. Measuring

Chelsea holds two buckets next to each other and says to Katia, "We should use the blue one, because it's bigger than the green one."

KDI 36. Measuring

The children line up the balls at the top of the hill; they speculate which one will get to the bottom first. The balls start rolling and Tony calls, "Look! The big red one is the fastest." Charlie says, "Yeah, because it's the biggest."

There are lots of patterns to observe in nature, including the patterns of petal growth on a flower.

KDI 36. Measuring

LaDwan and Veronica are looking through the loose planks for one they could use to make a bench. "How about this one?" asks Veronica. "No, too short," replies LaDwan, as she holds up another one. "What about this tall one?"

KDI 37. Unit

It is Henry's turn on the swing and he tells Aden, "Turn over the sand timer as soon I get on the swing to measure my turn, okay?"

Young children also notice mathematical patterns that occur in nature, for example, patterns in rocks, the pattern of petal growth on a flower, the pattern of rings that forms when you drop a pebble in a stream, and the pattern of bird or animal foot prints. These natural patterns are everywhere for children to observe and explore. Let's take a look and see what kinds of patterns children find in their outdoor environments.

KDI 38. Patterns

Julia rides her tricycle with a pinwheel taped to the handlebars. "Tansy," she calls to the teacher, "look at the pinwheel. It has a stripe pattern on two of the wings. See, it goes silver, red; silver, red; silver, red; silver, red."

KDI 38. Patterns

Zoe pokes random holes in the sand with a stick. When Alan (a teacher) sits down beside her, she turns to him and says, "Pattern."

KDI 38. Patterns

Some of the children talk about the garden they are going to plant. Keisha says, "I want some flowers too — a row that goes pink, purple; pink, purple; pink, purple.

Opportunities for preschool-aged children to conduct data analysis during outside time are not as common as you might think and are often confused with classification (KDI 46, under Science and Technology). Data analysis is all about different ways of representing — and then using — information (data). The following are a few examples of practicing data analysis outside.

KDI 39. Data analysis

It is time to go in for lunch. Alan (a teacher) asks all the children in Tansy's group to stand on the grass and all the children in his group to stand on the blacktop so they can be in groups when they pick up their lunches. Ramona stands on the grass because she is in Tansy's group.

KDI 39. Data analysis

Benjamin is shooting baskets by himself. After taking a few shots, he gets a piece of sidewalk chalk from the shed and draws a line going across and then another one going through the other line like this:

When Tansy asks him what the marks are, he points to one side and says, "Those are the baskets I made." He points to the other side and says, "I missed those."

Math in Action: Kicking Balls With Evan and Sandy

During outside time, preschool teacher Linda watches one of the children, Evan, kick a rubber ball across the grass. She says to him, "You kicked the ball really far. It went all the way to the climber." "Yeah, watch this one," Evan replies. "I can make it go all the way to the fence. That's really, really far" (KDIs 35, 36). *He then kicks the ball again toward the fence. "Wow!" Linda observes. "That time the ball went almost all the way across the playground!"*

"Watch me, Linda!" shouts Sandy, holding a ball. "I can kick it really, really far — to the moon" (KDI 36). *"Wow, all the way to the moon? That is a long way away," acknowledges Linda. She watches Sandy kick the ball and then says, "I'm going to try and kick my ball a little bit closer than the moon. I'm going to kick it toward the sandbox."*

"Let's see who can kick their ball the farthest" (KDI 36), *suggests Evan. "Okay," replies Sandy, and she kicks her ball hard across the grass. "Wait!" yells Evan. "We have to do it together so we can measure. Put your ball by mine"* (KDI 37), *he says to Sandy. Linda explains to Sandy, "Sandy, Evan wants you to put your ball close to his so we can measure which one goes farthest." Linda puts her ball on the ground a short distance in front of Evan's ball." "No! It has to go right here," says Evan, pointing to a spot right next to his ball. "They have to be the same"* (KDI 37). *Linda restates, "So we have to make sure the balls start at the same place in order to measure." Sandy retrieves her ball, and Linda helps her place it next to the others. "Now, one,*

Bikes give children the opportunity to use their large muscles as well as develop their spatial awareness.

two, three, kick" (KDI 31), calls Evan. The children kick their balls across the grass. "Mine's the farthest" (KDI 36), shouts Evan. "Then it's Linda. Sandy's ball didn't go the right way. It doesn't count" (KDI 35). Sandy says, "I want to do it again." Linda agrees, saying "Me too! How about you, Evan?"

Support strategies

- After observing Evan kick the ball, Linda introduces the distance word *far* to see whether Evan would be interested in the idea of distance as he played.

- Linda uses direction words (*to* the climber, *across* the playground, *toward* the sandbox) as she describes where she and Evan kick the balls.

- Linda supports Sandy's understanding of the word *far* (*far — to the moon*) by acknowledging that far means *a long way away*. Then she extends the idea of *far* by using the word *closer* as a comparison.

- Linda places her ball in front of Evan's ball to check his understanding that the balls must start at the same point in order to be measured. When Linda restates Evan's directions to place the balls next to each other ("So we have to make sure the balls start at the same place in order to measure"), she introduces a measuring procedure.

Math in Action: Brynna and Jordyn Get Speeding Tickets

Brynna and Jordyn ride the big wheels around the blacktop. "Whoa, you're going fast," says their teacher Carl. "Better be careful, or you'll get a speeding ticket." He smiles and starts to walk away. Brynna calls to him, "Hey, you be the police and pretend you give us speeding tickets because we're going too fast" (KDI 36). "Okay," replies Carl, "I'll have to get my radar detector out so I know how fast you are going." He holds up his hand as if it were a radar detector. "Drive safely, ladies, so I don't have to give you a ticket," he says in a pretend police officer voice.

The girls begin racing the big wheels, and Carl makes a siren sound and stops Jordyn. "Excuse me, ma'am. Do you realize that you were going 100 miles an hour and the speed limit is 10? I'm going to have to give you a ticket," Carl says. He takes a sticky note from his pocket, writes 100, and draws a circle around it with a diagonal slash through the circle (to represent the "no" symbol). He says, "This ticket says that you can't go 100. Now you have to pay the ticket. That will be five dollars, ma'am." Jordyn takes the ticket and pretends to pay Carl five dollars by hitting him on the hand five times as she counts, "One, two, three, four, five" (KDI 32). Then she says, "There!" and drives off. Carl picks up a piece of sidewalk chalk and writes the number 10 on the blacktop. He calls to Brynna and Jordyn and says, "There...I wrote the speed limit on the road so you will remember how fast to go!"

Next Brynna says, "Pretend that I'm going fifty-seven hundred miles an hour" (KDI 31). Carl promptly responds with a siren sound and says to Brynna, "Excuse me, miss, but do you know how fast you were going?" Brynna replies, "Why, yes, I was going fifty-seven hundred miles an hour" (KDI 31). "Well," says Carl, "the speed limit is 10." He points to the number 10 on the blacktop. "Fifty-seven hundred is a number way bigger than 10. You are going way over the speed limit! I'm going to have to give you a ticket." He pulls out his paper, writes the number 7 on it, and hands it to Brynna. "This is how much you have to pay," Carl says. Brynna looks at the ticket and says, "Seven, okay" (KDI 31). She slaps Carl's hand seven times. "There, that's seven (KDI 32). Now I have to go to the store." She speeds off again.

Support strategies

- Carl uses measurement terms ("Do you realize that you were going 100 miles an hour and the speed limit is 10?") to describe how *fast* Jordyn is riding.

- Carl sees the opportunity to add numerals into the play when the girls ask him to write speeding tickets, and again when he writes the numeral 10 on the blacktop.

- Carl invites Brynna to use number words to answer his question about how fast she was going. He then compares the number she says ("fifty-seven hundred") to the number 10 ("Fifty-seven hundred is a number way bigger than 10. You are going way over the speed limit!"), which supports Brynna's understanding that the number she chose was indeed a very large number.

- Carl writes the numeral 7 on paper to support Brynna's understanding of number symbols and encourage her to count as she pretends to pay seven dollars for her speeding ticket.

Math in Action: Turns on the Swing

Tommy and Abby are swinging on the two swings on the playground. Joey and Kobe sit on the side, watching the falling sand in a large sand timer and waiting for it to stop so they can take their turns (KDI 37). Once the sand stops, Joey announces, "Time's up; our turn" (KDI 37). Joey and Kobe start their turns on the swings, and Abby flips over the large sand timer (KDI 37).

In the meantime, Tommy brings out a smaller sand timer and watches the sand run through it (KDI 37). Soon the sand runs through the smaller timer and Tommy announces, "Time's up. Our turn" (KDI 37). Kobe protests, "Hey that's not fair. We only got a short time" (KDI 36).

"The sand is still going," says Joey as he points to Abby's larger sand timer. "Get off," Tommy insists. He calls Betsy, the teacher, who comes over to help. "What's the problem? You all look quite upset," Betsy observes. "They won't get off the swing" (KDI 35), says Tommy. "The sand timer is still going," explains Joey, pointing to the large timer Abby is holding. "We're using Abby's timer," says Joey. "No!" insists Tommy. "My timer is done." Kobe explains, "That's a small timer. You have to use the big timer. The big timer takes longer. The small timer is too short" (KDIs 36, 37).

Betsy listens and then restates Kobe's argument: "You're saying that the small timer that Tommy used has less time. The big timer takes more time for the sand to run through." "Yes," Kobe and Joey confirm in unison (KDI 36). "I wonder why that is," says Betsy. "The little timer has less sand in it," explains Abby, "so it doesn't take as long to go down" (KDI 36). "Ah," says Betsy, nodding her head and turning to Tommy. "Tommy, the little timer has less sand in it so it doesn't take as long to go through. Kobe and Joey want to use the big timer so they can have the same time as you and Abby had. Hmm, what do you think we should we do?" Betsy asks. "How about the little timer two times" (KDI 32)? Abby suggests. "Turn the little timer over twice?" asks Betsy. "I wonder if that would be enough time. Would you like to try it?" "Okay," they all agree.

As evident in this Mathematics in Action scenario, sand timers are a great way to help children understand the measurement of time and also — sometimes — avoid conflicts.

Outside Time

In addition to developing children's gross-motor skills, swinging helps children recognize spatial relationships among people and objects.

Support strategies

- Betsy listens to the children describe the problem and restates it using comparison words to help the children think about measuring time: "You're saying that the small timer that Tommy used has *less* time. The big timer takes *more* time for the sand to run through."

- Betsy makes the connection between the amount of sand in each timer and the length of time it takes the sand to run through: "Tommy, the little timer has less sand in it so it doesn't take as long to go through. Kobe and Joey want to use the big timer so they can have the same time as you and Abby had."

- Betsy supports Abby's idea of turning the little timer over two times, a solution that also satisfies Tommy, Kobe, and Joey.

Appendices

Appendix A: HighScope Key Developmental Indicators

Appendix B: HighScope Daily Routine

Appendix C: HighScope Preschool Wheel of Learning

Appendix A: HighScope Key Developmental Indicators

A. Approaches to Learning

1. **Initiative:** Children demonstrate initiative as they explore their world.
2. **Planning:** Children make plans and follow through on their intentions.
3. **Engagement:** Children focus on activities that interest them.
4. **Problem solving:** Children solve problems encountered in play.
5. **Use of resources:** Children gather information and formulate ideas about their world.
6. **Reflection:** Children reflect on their experiences.

B. Social and Emotional Development

7. **Self-identity:** Children have a positive self-identity.
8. **Sense of competence:** Children feel they are competent.
9. **Emotions:** Children recognize, label, and regulate their feelings.
10. **Empathy:** Children demonstrate empathy toward others.
11. **Community:** Children participate in the community of the classroom.
12. **Building relationships:** Children build relationships with other children and adults.
13. **Cooperative play:** Children engage in cooperative play.
14. **Moral development:** Children develop an internal sense of right and wrong.
15. **Conflict resolution:** Children resolve social conflicts.

C. Physical Development and Health

16. **Gross-motor skills:** Children demonstrate strength, flexibility, balance, and timing in using their large muscles.
17. **Fine-motor skills:** Children demonstrate dexterity and hand-eye coordination in using their small muscles.
18. **Body awareness:** Children know about their bodies and how to navigate them in space.
19. **Personal care:** Children carry out personal care routines on their own.
20. **Healthy behavior:** Children engage in healthy practices.

D. Language, Literacy, and Communication[1]

21. **Comprehension:** Children understand language.
22. **Speaking:** Children express themselves using language.
23. **Vocabulary:** Children understand and use a variety of words and phrases.
24. **Phonological awareness:** Children identify distinct sounds in spoken language.
25. **Alphabetic knowledge:** Children identify letter names and their sounds.
26. **Reading:** Children read for pleasure and information.
27. **Concepts about print:** Children demonstrate knowledge about environmental print.
28. **Book knowledge:** Children demonstrate knowledge about books.
29. **Writing:** Children write for many different purposes.
30. **English language learning:** (If applicable) Children use English and their home language(s) (including sign language).

---

[1] Language, Literacy, and Communication KDIs 21–29 may be used for the child's home language(s) as well as English. KDI 30 refers specifically to English language learning.

E. Mathematics

31. **Number words and symbols:** Children recognize and use number words and symbols.
32. **Counting:** Children count things.
33. **Part-whole relationships:** Children combine and separate quantities of objects.
34. **Shapes:** Children identify, name, and describe shapes.
35. **Spatial awareness:** Children recognize spatial relationships among people and objects.
36. **Measuring:** Children measure to describe, compare, and order things.
37. **Unit:** Children understand and use the concept of unit.
38. **Patterns:** Children identify, describe, copy, complete, and create patterns.
39. **Data analysis:** Children use information about quantity to draw conclusions, make decisions, and solve problems.

F. Creative Arts

40. **Art:** Children express and represent what they observe, think, imagine, and feel through two- and three-dimensional art.
41. **Music:** Children express and represent what they observe, think, imagine, and feel through music.
42. **Movement:** Children express and represent what they observe, think, imagine, and feel through movement.
43. **Pretend play:** Children express and represent what they observe, think, imagine, and feel through pretend play.
44. **Appreciating the arts:** Children appreciate the creative arts.

G. Science and Technology

45. **Observing:** Children observe the materials and processes in their environment.
46. **Classifying:** Children classify materials, actions, people, and events.
47. **Experimenting:** Children experiment to test their ideas.
48. **Predicting:** Children predict what they expect will happen.
49. **Drawing conclusions:** Children draw conclusions based on their experiences and observations.
50. **Communicating ideas:** Children communicate their ideas about the characteristics of things and how they work.
51. **Natural and physical world:** Children gather knowledge about the natural and physical world.
52. **Tools and technology:** Children explore and use tools and technology.

H. Social Studies

53. **Diversity:** Children understand that people have diverse characteristics, interests, and abilities.
54. **Community roles:** Children recognize that people have different roles and functions in the community.
55. **Decision making:** Children participate in making classroom decisions.
56. **Geography:** Children recognize and interpret features and locations in their environment.
57. **History:** Children understand past, present, and future.
58. **Ecology:** Children understand the importance of taking care of their environment.

Appendix B: HighScope Daily Routine

The components of HighScope's daily routine help to provide a consistent, balanced experience for children. At the end of each day, the teachers meet together to discuss what they observed and learned about individual children. They share and record anecdotes, and plan for the next day's activities based on what they learned about children that day.

Greeting time (5–10 minutes)
Greeting time provides a smooth transition from home to school. Teachers greet children, connect with parents, and read books in a cozy setting. The message board gives children and teachers a chance to share important information for the day. Parents often join their children for this part of our day.

Planning time (10–15 minutes)
In their small groups, children indicate what they choose to do during work time (typically what they will do first). Their teacher will try to understand children's plans and often try to help children extend their plans.

Work time (45–60 minutes)
Children carry out their initial and subsequent plans. Children can work with any of the materials in any of the interest areas. Teachers observe children and look for opportunities to enter into children's activities to encourage their thinking, extend their play, and help them wrestle with problem-solving situations.

Cleanup time (10 minutes)
Children and teachers together return materials and equipment to their storage spaces and, when appropriate, put away or find display space for their personal creations.

Recall time (10–15 minutes)
Recall time brings closure to the planning, work time, recall sequence. In their small group, children reflect on, talk about, and/or show what they have done at work time.

Meal- and snacktimes (20 minutes)
Children and teachers share nutritious food and interesting conversation together in a relaxed, family-style manner.

Large-group time (10–15 minutes)
Children and teachers gather together to play games, tell and reenact stories, sing songs, do fingerplays, dance, play musical instruments, or reenact special events. This time is an opportunity for each child to participate in a large group, sharing ideas and learning from the ideas of others.

Small-group time (15–20 minutes)
Each teacher meets with their consistent small group of children to work with materials planned and introduced by the teacher. Although the teacher chooses and introduces the materials, each child has control over what he or she will do with them.

Outside time (30–40 minutes)
Children engage in vigorous, noisy outdoor play. Teachers participate in, and support, children's play outdoors.

Appendix C: HighScope Preschool Wheel of Learning

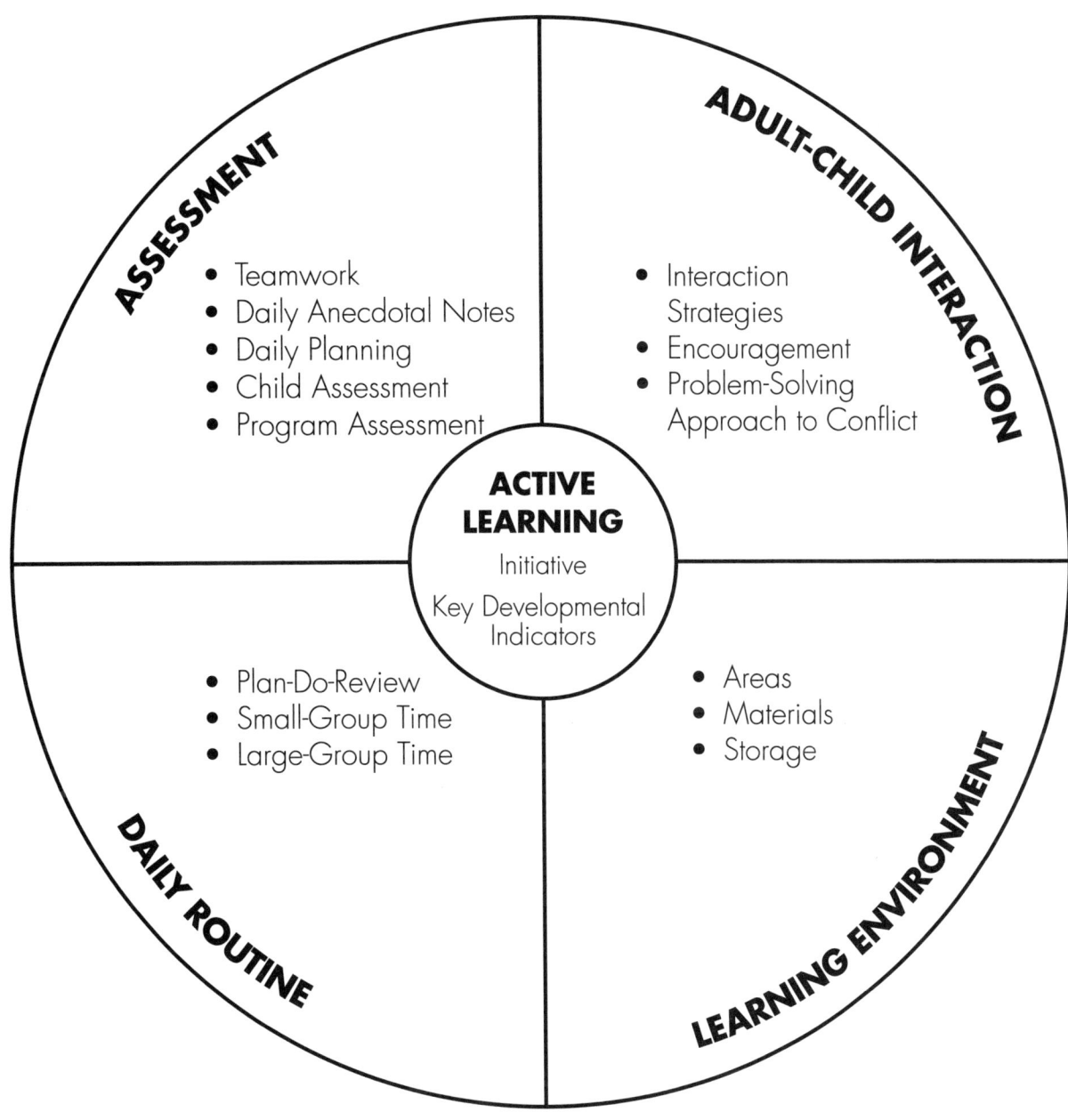

References

Arithmetic. (2013). In *Merriam-Webster's online dictionary* (11th ed.). Retrieved from http://www.merriam-webster.com/dictionary/arithmetic.

Beatles, The. (2009). Come together. *Abbey road* [Audio CD]. London, United Kingdom: EMI Music Publishing. (Original work published 1969)

Boisvert, C., & Gainsley, S. (2006). *50 large-group activities for active learners*. Ypsilanti, MI: HighScope Press.

Bradberry, T., & Greaves, J. (2009). *Emotional intelligence 2.0*. San Diego, CA: Talent Smart.

Clements, D. (2001, January). Mathematics in preschool. *Teaching Children Mathematics, 7*(5), 270–278.

Dawkins, R. (1996). *The blind watchmaker: Why the evidence of evolution reveals a universe without design*. New York: W.W. Norton & Company.

Duckworth, E. (2006). *The having of wonderful ideas: And other essays on teaching and learning* (3rd ed.). New York: Teachers College Press.

Duncan, G. J., Dowsett, C. J., Claessens, A., Magnuson, K., Huston, A. C., Klebanov, P.,...Japel, C. (2007). School readiness and later achievement. *Developmental Psychology, 43*(6), 1428–1446. http://dx.doi.org/10.1037/0012-1649.43.6.1428

Eisenberg, N. (Vol. Ed.). (2006). *Handbook of child psychology: Vol. 3. Social, emotional, and personality development* (6th ed.). Hoboken, New Jersey: Wiley.

Epstein, A. S. (2003). How planning and reflection develop young children's thinking skills. *Young Children* on the Web. Retrieved from http://www.naeyc.org/yc/pastissues/2003/september

Epstein, A. S. (2012). *Mathematics*. Ypsilanti, MI: HighScope Press.

Epstein, A. S., & Gainsley, S. (2010). *"I'm older than you. I'm five!" Math in the preschool classroom* (2nd ed.). Ypsilanti, MI: HighScope Press.

Epstein, A. S., & Hohmann, M. (2012). *The HighScope Preschool Curriculum*. Ypsilanti, MI: HighScope Press.

Evans, B. (2007). *"I know what's next!" Preschool transitions without tears or turmoil*. Ypsilanti, MI: HighScope Press.

Fulghum, R. (1990). *All I really need to know I learned in kindergarten*. New York: Villard Books.

Fuson, K. C., Clements, D. H., & Beckmann, S. (2010). *Focus in prekindergarten: Teaching with curriculum focal points*. Reston, VA: National Council for Teachers of Mathematics & Washington, DC: National Association for the Education of Young Children.

Gainsley, S. (2005). Cleanup time strategies from the Demonstration Preschool. In N. A. Brickman, H. Barton, & J. Burd (Eds.), *Supporting young learners* (Vol. 4, pp. 91-94). Ypsilanti, MI: HighScope Press.*

Gainsley, S. (2008). *From message to meaning: Using a daily message board in the preschool classroom*. Ypsilanti, MI: HighScope Press.

Gainsley, S. (2011). Look, listen, touch, feel, taste: The importance of sensory play. *Extensions, 25*(5), 1–12. Available at the HighScope *Extensions* archive at highscope.org.

Geist, E. (2009). *Children are born mathematicians*. Upper Saddle River, NJ: Pearson Education.

Gresham, G. (2007). A study of mathematics anxiety in pre-service teachers. *Early Childhood Education Journal, 35*(2), 181–188. http://dx.doi.org/10.1007/s10643-007-0174-7

Gross, F. E. (2006). Math talk with young children: One parent's experience. In B. Neugebauer (Ed.), *Curriculum: Brain research, math, science* (pp. 47–49). Redmond, WA: Exchange Press.

Handler, D., & Epstein, A. S. (2010). Nature education in preschool. *Extensions, 25*(2), 1–7. Available at the HighScope *Extensions* archive at highscope.org.

Harris, V. (1996). Open-air learning experiences. In N. Brickman (Ed.), *Supporting young learners* (Vol. 2, pp. 119–126). Ypsilanti, MI: HighScope Press.*

Hendrick, J., & Weissman, P. (2010). *Total learning: Developmental curriculum for the young child* (8th ed.). Upper Saddle River, NJ: Pearson Education.

---

*Also available at the HighScope *Extensions* archive at highscope.org.

HighScope Early Childhood Education Staff. (2009). *Small-group times to scaffold early learning.* Ypsilanti, MI: HighScope Press.

HighScope Educational Research Foundation. (2009). *Numbers Plus preschool mathematics curriculum.* Ypsilanti, MI: HighScope Press.

Hohmann, M., Weikart, D. P., & Epstein, A. S. (2008). *Educating young children* (3rd ed.). Ypsilanti, MI: HighScope Press.

Holt, J. (1983). *How children learn* (Rev. ed.). Cambridge, MA: Da Capo Press. (Original work published 1967)

Holt, J. (2008, March 3). Numbers guy: Are our brains wired for math? *The New Yorker.* Retrieved from http://www.newyorker.com/reporting/2008/03/03/080303fa_fact_holt?currentPage=all

Hynes-Berry, M., & Itzkowich, R. (2009). The gift of error. In C. Gibbs & A. Gibbons (Eds.), *Conversations on early childhood teacher education* (pp. 104–112). Auckland, New Zealand: New Zealand Tertiary College.

Illich, I. (1971). Deschooling society. New York: Harper & Row.

Klein, A., Starkey P., Molfese, V., Brown, E. T., & Molfese, D. (2008). *Stages in the development of a mathematics intervention for public preschool programs.* Retrieved from ERIC database. (ED525313)

Marcon, R. A. (2002). Moving up the grades: Relationship between preschool model and later school success. *Early Childhood Research and Practice, 4*(1). Retrieved from http://ecrp.uiuc.edu/v4n1/marcon.html

Marshall, B. (2009). Scaffolding children's learning at small-group time. *Extensions, 23*(4), 1–3. Available at the HighScope *Extensions* archive at highscope.org.

Marshall, B. (with Lockhart, S., & Fewson, M.). (2007). *HighScope step by step: Lesson plans for the first 30 days.* Ypsilanti, MI: HighScope Press.

Moomaw, S. (2011). *Teaching mathematics in early childhood.* Baltimore, MD: Paul H. Brookes.

National Association for the Education of Young Children (NAEYC) & National Council of Teachers of Mathematics (NCTM). (2009). *Where we stand on early childhood mathematics.* Retrieved from http://www.naeyc.org/files/naeyc/file/positions/ecmath.pdf

National Association for the Education of Young Children (NAEYC) & National Council of Teachers of Mathematics (NCTM). (2010). *Early childhood mathematics: Promoting good beginnings — Joint position statement.* Retrieved from https://www.naeyc.org/files/naeyc/file/positions/psmath.pdf

National Council of Teachers of Mathematics (NCTM). (2006). *Curriculum focal points for prekindergarten through grade 8 mathematics: A quest for coherence.* Reston, VA: Author.

National Research Council (NRC). (2000). *Eager to learn: Educating our preschoolers.* Washington, DC: National Academies Press.

National Research Council (NRC). (2009). *Mathematics learning in early childhood: Paths toward excellence and equity.* Washington, DC: National Academies Press.

Scaffolding. (2011). In *American Heritage dictionary of the English language* (5th ed.). Boston: Houghton Mifflin Harcourt.

Siegler, R. S. (2009). Improving numerical understanding of children from low-income families. *Child Development Perspectives, 3*(2), 118–124. http://dx.doi.org/10.1111/j.1750-8606.2009.00090.x

Sorooshian, P. (2009, February 5). Getting past your own math anxiety. [Web log post]. Retrieved from http://learninghappens.wordpress.com/category/math/math-anxiety-math/

Vogel, N. (2001). *Making the most of plan-do-review.* Ypsilanti, MI: HighScope Press.

White, E. B. (1952). *Charlotte's web.* New York: HarperCollins.

About the Authors

Polly P. Neill is the author of *Real Science in Preschool: Here, There, and Everywhere* and has contributed to several other HighScope books. Polly collaborated in the development and design of many of HighScope's online course offerings. She facilitates online training as well as Demonstration Preschool observation sessions.

Suzanne Gainsley is the director of the HighScope Demonstration Preschool and is a HighScope early childhood specialist. She is the author of *From Message to Meaning: Using a Daily Message Board in the Preschool Classroom;* coauthor of two books in the Teachers' Idea Series, including *"I'm Older Than You. I'm Five!" Math in the Preschool Classroom* and *50 Large-Group Activities for Active Learners;* and coauthor of *Preschool Activities for Family Child Care* and *Activities for Home Visits: Partnering With Preschool Families.*